# ANIMAL EXPERIMENTATION

Other books in the At Issue series:

AIDS in Developing Countries
Alcohol Abuse
The Attack on America: September 11, 2001
Bilingual Education
Bulimia
The Central Intelligence Agency
Child Sexual Abuse in the Catholic Church
Cloning
Computers and Education
Creationism vs. Evolution
Date Rape
Does Capital Punishment Deter Crime?
Drunk Driving
Fighting Bioterrorism
Food-Borne Illnesses
Foreign Oil Dependence
Genetically Engineered Foods
Guns and Crime
Homeland Security
Home Schooling
Is Global Warming a Threat?
Islamic Fundamentalism
Is Media Violence a Problem?
Is Military Action Justified Against Nations That Support Terrorism?
Is the Death Penalty Fair?
Marijuana
Missile Defense
National Security
Organ Transplants
Performance-Enhancing Drugs
Police Corruption
Reality TV
Reparations for American Slavery
School Shootings
Should Abortion Rights Be Restricted?
Should There Be Limits to Free Speech?
Slavery Today
Teen Smoking
U.S. Policy Toward Rogue Nations
Vaccinations
Video Games
Violent Children
White Supremacy Groups

# ANIMAL EXPERIMENTATION

Cindy Mur, *Book Editor*

Bonnie Szumski, *Publisher*
Scott Barbour, *Managing Editor*
Helen Cothran, *Senior Editor*

San Diego • Detroit • New York • San Francisco • Cleveland
New Haven, Conn. • Waterville, Maine • London • Munich

LIBRARY OF CONGRESS CATALOGING-IN-PUBLICATION DATA

Animal experimentation / Cindy Mur, book editor.
   p. cm. — (At issue)
Includes bibliographical references and index.
ISBN 0-7377-1999-0 (lib. bdg. : alk. paper) —
ISBN 0-7377-2000-X (pbk. : alk. paper)
   1. Animal experimentation—Moral and ethical aspects. 2. Animal rights.
3. Laboratory animals. I. Mur, Cindy. II. At issue (San Diego, Calif.)
HV4915.A634 2004
179'.4—dc22                                      2003047237

Printed in the United States of America

# Contents

Page

# Introduction

Working for the periodical *Animal's Agenda*, Rick Bogle discovered disturbing aspects of research projects on nonhuman primates (henceforth referred to as primates). He found one researcher who was "depriving infant rhesus macaques of key nutrients and stud[ying] the results, such as chronic diarrhea and neural impairment," and another researcher who was "learning how to bolt the heads of three-month-old monkeys into a restraint device and inject[ing] chemicals into their brains to induce seizures." In other experiments, baby monkeys were separated from their mothers so researchers could study conditions like depression, aggression, and mother-infant bonding.

Torturous experiments on primates, like those depicted above, make most people uncomfortable. For many years in the United States, heated controversy has surrounded animal experimentation in general, but no issue is more emotionally charged than using primates in medical tests. At an emotional level, humans recognize something of themselves in primates, and they are therefore reluctant to approve the use of primates for experimentation, especially if the test would be painful. Polls indicate that the public believes a difference exists between primates and other animals and that primates have much in common with humans.

These feelings of kinship drive animal protection groups to prevent experimentation on primates. However, those involved with research on primates argue that primate experiments are necessary to find cures for human diseases. The debate over primate testing centers around two issues: the effectiveness of testing on primates and the ethical questions raised when using humanity's closest living relatives for experimentation.

Researchers and animal rights activists disagree on the medical contributions of primate testing. Scientists assert that animal research in general, and primate research specifically, has been vital to protecting human health. According to the Scientific Steering Committee for the European Commission, "Experiments on live animals are powerful ways of better understanding the complex biological mechanisms" of the human body. Scientists use primates whose immune systems are similar to humans to make sure that vaccines are safe, for example. The committee members believe that trials for AIDS, malaria, tuberculosis, hepatitis C, and immune-based diseases depend upon primate testing. Neural testing on primates has led to advances in the treatment of Parkinson's disease. These advances were made possible by the fact that humans and primates are remarkably similar.

Indeed, scientific data indicate that 97.7 percent of the DNA in apes and humans is the same. Chimpanzee DNA matches 98.7 percent of human DNA. (Most mammals have DNA structures that match human genes by at least 90 percent.) While acknowledging that genetic similarities between primates and humans exists, opponents to testing, like C. Ray Greek, a medical doctor and author of several books attacking the efficacy of animal experiments, dismiss the idea of physiological resemblance. "The primate brain is not a scaled-down version of our brain," Greek told *New Scientist.* "Chimp brains and human brains are similar in structure, but that doesn't mean they perform the same functions."

Greek's statement is at the crux of the arguments presented by animal rights advocates. Advocates believe that although primates exhibit humanlike qualities, their physiology makes them poor test subjects. A statement from the European Coalition to End Animal Experiments outlines why this is the case:

> After decades of research on primates, scientists have repeatedly failed to make significant breakthroughs in fully understanding the onset and progression of HIV or AIDS, cot death [Sudden Infant Death Syndrome], epilepsy, Parkinson's or Alzheimer's disease, or cancer—all human conditions which have been thoroughly, though pointlessly, explored through research on primates. The fundamental flaw underlying the research of human diseases in primates is that researchers can only artificially recreate the symptoms of human diseases in primates, which is very different from studying a naturally occurring disease in a biologically relevant animal such as a human patient.

Ongoing AIDS experiments illustrate the problems with using primates for testing. The Physicians Committee for Responsible Medicine (PCRM), a group of doctors that promotes alternatives to animal experimentation, describes the pitfalls of using chimpanzees in AIDS research:

> None have become clinically ill, in spite of being infected with several different strains of the virus, having their immune systems altered with drugs, having treatments designed to specifically destroy the cells which are thought to be most active in protecting the body from HIV infection, and being co-infected with other viruses which were presumed to help HIV gain a foothold. Experimenters have even injected human HIV-infected brain tissue directly into chimpanzee brains, but to no avail.

PCRM members believe that using primates for AIDS research wastes money and time that could be better spent on more effective means of testing, such as clinical trials or in-vitro experiments. These experiments use human rather than animal subjects or cell tissue. Animal rights activists believe that animal experiments harm humanity by taking resources from these more effective techniques.

The value of primate experiments is just one subject of contention between researchers and animal rights activists. These two groups also disagree on whether it is ethical to experiment on primates.

The general arguments for and against the ethical use of any animal for experimentation are important to understand because they provide a foundation for the debate over primate testing. On one hand, animal rights activists believe that all animals deserve the same rights as humans, including the right to freedom from unnecessary or unjustifiable pain or discomfort. Tom Regan, a professor of philosophy at North Carolina State University and a leader in the field of animal rights, suggests that all living beings have an inherent value and that to use any animals for experimentation is evil. He believes that, as moral agents—those who have the ability to apply moral principles in decision making—humanity has a duty to practice that morality not just on other moral agents but on moral patients as well—those who cannot apply moral principles, such as children, the mentally disabled, and animals—even though these beings cannot reciprocate.

Researchers, on the other hand, believe that their experiments on animals are morally justified. They assert that animal experimentation in general has benefited humanity. They argue that the advantages for humankind outweigh the harm done to animals.

Many advocates of animal testing also do not believe that animals and humans are moral equivalents; therefore, they do not think that animals deserve the same rights as humans. Tibor Machan, a philosophy professor at Chapman University in Orange, California, believes "such rights would only arise if animals developed into moral agents, which they haven't . . . no one is expecting animals to be kind, compassionate, considerate of their own victims." Many advocates of animal experimentation feel that the fact that humans can feel guilt over experiments demonstrates they are superior to all animals.

Some supporters of animal experimentation hold the belief that God placed animals on Earth for the benefit of humankind, and therefore humans have the right and obligation to use animals as needed. These supporters assert that humans are made in the image of God, so to equate them to any other animal degrades humankind. According to David R. Carlin, a professor of philosophy and sociology at the Community College of Rhode Island, people hold a special place in the universe. Carlin writes, "To reduce human nature to nothing more than its biological status is to attack this ancient and exalted conception of human nature." Other supporters of animal experimentation go further, saying that not only is it not harmful for humanity to experiment on animals, it is ethically wrong not to perform those experiments if people will benefit.

The debate over animal experimentation is made more contentious when specifically considering primates because they exhibit many human-like characteristics. Animal rights advocate Rick Bogle supports his arguments by citing data that indicate "just how cognitively sophisticated and emotionally sensitive monkeys and apes are." He points out that apes have demonstrated the ability to use sign language at a level equal to that of a three- or four-year-old child as well as "joke, lie and empathize with humans and other animals." When provided with a mirror, apes will examine and groom themselves, demonstrating a sense of self similar to humans. One gorilla, Koko, has scored between seventy and ninety-five on human IQ tests; the average human has an IQ of one hundred.

Steven Wise, an animal rights lawyer and a vocal voice for the rights of primates, seeks not only to prevent experimentation but to also provide primates with rights equal to humans so that they would no longer be considered property, to be used as human owners see fit. In fact, Wise compares the plight of primates to that of human slaves—sentient beings without the "rights of bodily integrity and bodily liberty." He believes it is unfair for an intelligent, feeling primate to have no more rights than a chair when an encephalic child who has no brain has the same rights as any human. Wise challenges the idea that humans are intrinsically more valuable than primates and in so doing, calls into question the notion that humans are superior to all animals.

Under pressure from animal rights advocates such as Wise, some countries have taken dramatic strides to halt primate experiments. New Zealand has already provided great apes (chimpanzees, orangutans, and gorillas) with legal rights that protect them from being used for research, testing, or teaching. The United Kingdom has banned all experiments using great apes; the nation's leaders believe that such procedures cannot be justified because apes have too high a level of sentiency.

Animal rights activists around the world continue to advance legislation that will ensure that primates are no longer subject to the agonizing tests described by Bogle. However, they meet resistance from researchers who view primate testing as necessary for the development of drugs and vaccines for certain diseases. In *At Issue: Animal Experimentation*, activists, scientists, researchers, and educators debate the issues surrounding animal testing. The controversy over the role of animals in medicine will likely persist as long as some diseases remain uncured.

# 1

# Animals Are Entitled to Rights

## Tom Regan

*Tom Regan is a professor of philosophy at North Carolina State University in Raleigh and has written a number of books on the subject of animal rights including* Defending Animal Rights *and* The Animal Rights Debate *from which this viewpoint is taken.*

Humans do not have the right to use animals for their own ends; therefore, animals should not be used for scientific experimentation. Some critics of this view raise questions about the scope of animal rights, such as whether they would have the right to vote. These critics believe that since animals cannot respect our rights, we should not grant rights to them. In fact, all animals have the right to be treated with respect. Like children, they do not have to vote or respect another's rights to have rights themselves. To continue to deliberately and cruelly violate the rights of animals exemplifies evil.

M any people resist the idea of animal rights. Some of the objections are raised by academic philosophers; for example, some question the cogency of attributing a unified, complicated psychology to animals who are unable to use a language. Other objections are the stuff of everyday incredulity; objections of this type are voiced not only by philosophers but also by skeptical members of the general public. . . . Here I limit myself to answering some of the most common objections of the second type.

## The absurdity of animal rights

Some critics challenge the idea of animal rights head-on. If animals have rights, they contend, we will have to acknowledge their right to vote, marry, and file for divorce, all of which is absurd. Thus, animals have no rights.

Now, part of what is said is true: any view that entails that animals have the right to vote, marry, and file for divorce is absurd. Clearly, how-

Tom Regan, *The Animal Rights Debate*. Lanham, MD: Rowman & Littlefield Publishers, Inc., 2001. Copyright © 2001 by Rowman & Littlefield Publishers, Inc. Reproduced by permission.

ever, the rights view entails nothing of the sort. Different individuals do not have to have all of the same rights in order to have some of the same rights. An eight-month-old child, for example, does not have either the right to vote or the other rights enumerated in the objection. But this does not mean that the child lacks the right to be treated with respect. On the contrary, young children possess this right, at least according to the rights view. And since these children possess this right while lacking the rights mentioned in the objection, there is no reason to judge the status of animals differently. Animals need not have the right to vote, marry, or file for divorce, if they have the right to be treated with respect.

## No reciprocity

Critics of animal rights sometimes maintain that animals cannot have rights because animals do not respect human rights. Again, part of this objection is correct: animals do not respect our rights. Indeed, animals (we have every good reason to believe) have no idea of what it even means to respect someone else's rights. However, this lack of understanding and its behavioral consequence (namely, the absence of animal behavior that exhibits respect for human rights) do not undermine attributing rights to animals.

Once again, the moral status of young children should serve to remind us of how unfounded the requirement of reciprocity is. We do not suppose that young children must first respect our rights before we are duty bound to respect theirs. Reciprocity is not required in their case. We have no nonarbitrary, nonprejudicial reason to demand that animals conform to a different standard.

## Line drawing

"But where do you draw the line? How do you know exactly which animals are subjects-of-a-life[1] (and thus have a right to be treated with respect) and which animals are not?" There is an honest, simple answer to these vexing questions: we do not know exactly where to draw the line. Consciousness, which is presupposed by those who are subject-of-a-life, is one of life's great mysteries. Whether mental states are identical with brain states or not, we have massive evidence that our having any mental states at all presupposes our having an intact, functioning central nervous system and brain activity above the brain stem. Where exactly this physiological basis for consciousness emerges on the phylogenic scale, where exactly it disappears, no one can really know with certainty. But neither do we need to know this.

We do not need to know exactly how tall a person must be to be tall, before we can know that Shaquille O'Neal is tall. We do not need to know exactly how old a person must be to be old, before we can know that Grandma Moses was old. Similarly, we do not need to know exactly where an animal must be located on the phylogenic scale to be a subject-of-a-life, before we can know that the animals who concern us—those who are raised to be eaten, those who are ranched or trapped for their fur, or those

---

1. any being that has memory, perception, desire, and an understanding of its future

who are used as models of human disease, for example—are subjects-of-a-life. We do not need to know everything before we can know something. Our ignorance about how far down the phylogenic scale we should go before we say that consciousness vanishes should not prevent us from saying where it is obviously present. . . .

## Only humans are inherently valuable

Other objections to animal rights take different forms. For example, some critics maintain that because all and only human beings have inherent value, all and only human beings have a right to be treated with respect. How might this view be defended? Shall we say that all and only humans have the same level of intelligence, or autonomy, or reason? But there are many humans who lack these capacities and yet who, according to the rights view, have value above and beyond their possible usefulness to others. Will it then be suggested that this is true only in the case of human beings because only humans belong to the right species, the species *Homo sapiens?* But this is blatant speciesism.

## Animals have less inherent value

Some critics contend that while animals have some inherent value, they have less, even far less, than we do. Attempts to defend this view can be shown to lack rational justification. What could be the basis of our having more inherent value than animals? Their lack of reason, or autonomy, or intellect? Only if, as is true of moral elitists like Aristotle, we are willing to make the same judgment in the case of humans who are similarly deficient. But it is not true that human subjects-of-a-life who have significantly less mental ability than is normal therefore have less inherent value than we do. It is not true (at least it is not true according to the rights view) that these humans may be treated merely as means in cases where it would be wrong to treat more competent humans in the same way. Those humans who are less mentally endowed are not the natural slaves of those of us who, without our having done anything to deserve it, are more fortunate when it comes to our innate intelligence. That being so, we cannot rationally sustain the view that animals like these humans in the relevant respects have less inherent value. All who have inherent value have it equally, all who exist as subjects-of-a-life have the same morally significant value—whether they be human animals or not.

## Only humans have souls

Some people think that the crucial difference between humans and other animals is that we do, whereas they do not, have a soul. After all, we are the ones who are created "in the image of God"; *that* is why all humans have inherent value and why every nonhuman animal lacks value of this kind. Proponents of this view have their work cut out for them. I am myself not ill disposed to the proposition that there are immortal souls. Personally, I profoundly hope I have one. But I would not want to rest my position on a controversial issue like this one about inherent value, on the even more controversial question about who or what has an immortal soul. Rationally, it is better to resolve moral issues without making more controversial assumptions than are needed. The question of who has inherent value is such a question, one that is resolved more rationally by reference to the subject-of-a-life criterion, without the introduction of the idea of immortal souls, than by its use.

---

*Animals need not have the right to vote, marry, or file for divorce, if they have the right to be treated with respect.*

---

But suppose we grant, for the sake of argument, that every human has an immortal soul and that every other animal lacks one. Would this justify the way we treat animals? More specifically, would this justify using mice in LD50[2] tests, or raising calves after the fashion of the milk-fed veal trade? Certainly not. Indeed, if anything, the absence of a soul arguably makes such conduct even more reprehensible than it already is. For consider: If we have immortal souls, then however bad our earthly lives have been, however much suffering and personal tragedy we have had to endure, we at least can look forward to the prospect of having a joyful existence in the eternal hereafter. Not so a milk-fed veal calf or a mouse whose internal organs burst in response to heavy doses of paint stripper. Absent a soul, there can be no other life after this one that compensates them for their misery while on Earth. Denied the possibility of such compensation, which we are assuming all humans enjoy, the pain, loneliness, terror, and other evils these animals suffer are, if anything, arguably worse than those experienced by human beings. So, no, the soul argument will not serve the purposes of those seeking a justification of the tyranny humans exercise over other animals. Just the opposite.

## What about plants?

Inherent value, according to the rights view, belongs equally to all those who are subjects-of-a-life. Whether it belongs to other forms of life, including plants, or even to rocks and rivers, ecosystems, and the biosphere, are questions the rights view leaves open for others to explore, noting only that the onus of proof will be on those who wish to attribute inherent value beyond subjects-of-a-life to offer a principled, nonarbitrary, nonprejudicial, and rational defense of doing so.

Wherever the truth might lie concerning these matters, the rights view's implications concerning the treatment of animals are unaffected. We do not need to know how many people are eligible to vote in the next presidential election before we can know whether we are. Why should we need to know whether plants and the biosphere are inherently valuable before we can know that animals are?

And we do know that the billions of animals that, in our culture, are routinely eaten, trapped, and used in laboratories, for example, are like us in being subjects-of-a-life. And since, to arrive at the best account of our

---

2. a test in which chemicals are administered until half the subjects die

duties to one another, we must recognize *our* equal inherent value and *our* equal right to be treated with respect, reason—not mere sentiment, not unexamined emotion, but reason—compels us to recognize *their* equal inherent value and *their* equal right to respectful treatment.

## The magnitude of evil

Whether the ways animals are treated by humans adds to the evil of the world depends not only on how they are treated but also on what their moral status is. Not surprisingly, the rights view represents the world as containing far more evil than it is customary to acknowledge. First, and most obviously, there is the evil associated with the ordinary, day-to-day treatment to which literally billions of animals are subjected. . . . If it is true . . . that these animals have a right to be treated with respect, then the massive, day-to-day invasion of their bodies, denial of their basic liberties, and destruction of their very lives suggest a magnitude of evil so vast that, like light-years in astronomy, it is all but incomprehensible.

---

*All who have inherent value have it equally . . . whether they be human animals or not.*

---

But this is not the end of the matter. For the magnitude of evil is much greater than the sum of the violations of animal rights and the morally wrong assaults on their independent value these violations represent. . . . One of the weaknesses of preference utilitarianism[3] is that it cannot rule out counting evil preferences in the process of reaching a fully informed judgment of moral right and wrong. This is a weakness that any plausible moral outlook must remedy, and the rights view has a way of doing so. . . . According to the rights view evil preferences are those preferences which, if acted on, either lead agents to violate someone's rights or cause others to approve of, or tolerate, such violations.

From the perspective of the rights view, therefore, the magnitude of the evil in the world is not represented only by the evil done to animals when their rights are violated; it includes as well the innumerable human preferences that are satisfied by doing so. That the majority of people who act on such preferences (e.g., people who earn a living in the fur industry or those who frequent KFC) do not recognize the preferences that motivate them as evil—indeed, that some will adamantly assert that nothing could be further from the truth—settles nothing. Whether the preferences we act on are evil is not something to be established by asking how strenuously we deny that they are; their moral status depends on whether by acting on them we are party to or complicit in the violation of someone's rights.

Are all those who act on evil preferences evil people? Not at all. . . . People are evil (at least this is the clearest example of what we mean) when their general character leads them to habitually violate others'

---

3. A moral theory based on the belief that the satisfaction of people's preferences is inherently good and that, therefore, the rightness of an action depends on how closely it satisfies those preferences.

rights *and* to do so cruelly, either by taking pleasure in or by feeling nothing (being indifferent) about the suffering or loss caused by the violation. While some who benefit from animal rights' violations may meet this description, the majority of people, including those who, as part of their day-to-day life, are supportive or tolerant of this evil, are not. In the vast majority of cases, I believe, those associated with the meat industry, for example, and those who support it by acting on their gustatory preferences, are not evil people. And the same is true of the vast majority of other people who either are themselves actively engaged in industries that routinely violate the rights of animals or are supportive or complicit in these violations.

---

*Animal rights increasingly is accepted as an appropriate moral norm.*

---

The judgment that otherwise decent people act on evil preferences in these ways may invite anger and resentment from some, hoots of derisive laughter from others; but it may also awaken still others to a larger sense of the moral significance of our life, including (even) the moral significance of our most mundane choices: what we put in our mouths and wear on our backs. Imperfect creatures that we are, living in an imperfect world, no one of us can be entirely free from our role in the evil around us. That recognition of the rights of animals reveals far more evil than was previously suspected is no reason to deny the magnitude of the evil that exists in the world at large or how much, on close examination, we find in ourselves; rather, our common moral task is to conscientiously search for ways to lessen both.

## The grounds of hope

How has it come to pass that people who genuinely care about animals, companion animals in particular, nonetheless find themselves supporting practices that are evil not only in their result but also in their origin? This is a question to give the most ardent animal rights advocate pause. Certainly I do not have an answer ready at hand. In fact, recent work by sociologists studying human attitudes and behavior suggests that animal rights is not an idea whose time has come.

In their studies of diverse human populations, Arnold Arluke and Clinton R. Sanders cite many of the "conflicts" and "contradictions" that characterize human-animal interactions. Do these "conflicts" and "contradictions" bother people? Hardly ever, according to the authors. Write Arluke and Sanders: "While inconsistency does occasionally come into an individual's awareness as a glaring problem calling for correction, most of the time, most people live comfortably with contradictions as a natural and normal part of everyday life." And, again: "[Living with contradictions] is not troublesome for ordinary persons because commonsense is not constrained to be consistent." For the great mass of humanity, then, loving animals and eating them, or respecting animals and wearing them, are not matters to lose any sleep over.

History suggests that humans are made of sturdier stuff. If most of us, most of the time, really had no trouble living with contradictions, slavery would still be with us and women would still be campaigning for the vote. While some people some of the time may be able to live with some contradictions, some inconsistencies, there must be thresholds above which the daily business of living is affected. . . . I clearly remember when this happened in my life. My reading of Gandhi awakened me to the realization that I held inconsistent beliefs and attitudes about unnecessary violence to human beings, on the one hand, and unnecessary violence to animal beings, on the other. And the death of a canine friend led me to the realization that I was placing some animals (dogs and cats, in particular) in one moral category and other animals (e.g., hogs and calves) in another, even as I realized that, when viewed in terms of their individual capabilities, there really was no morally relevant difference between them. I have no reason to believe my wanting to craft a coherent set of values for my life makes me any different from anyone else. None of us is so acculturated that we sleep-walk through our moral life. We know a contradiction in our values when we see one. If too few of us today are seriously troubled by our contradictory beliefs and attitudes toward animals, that may be because too few of us recognize where and why our beliefs and attitudes are contradictory. What is invisible must first be made visible before it can be seen; contradictions must first be seen before they can be honestly and directly addressed. One of the present essay's central purposes has been to help make some things more visible than before.

## Life changes

Evidence suggests that more and more people are beginning to come to terms with such inconsistencies and are changing their lives as a result. Take the fur industry, for example. As recently as the mid-1980s, seventeen million animals were trapped for their fur in the United States; by the early 1990s, that number was approximately ten million; for 1997–98, the total had fallen to four million and estimates for 1998–99 place the total at half that of the previous year. During this same period the number of caged-mink "ranches" declined from 1,000 to 401. In 1988, active trappers numbered 330,000; by 1994, there were fewer than half that number. And while there were almost eight hundred fur manufacturers in America in 1972, their ranks had dwindled to just over two hundred in 1992. Arizona, California, Colorado, Florida, Massachusetts, New Jersey, and Rhode Island have joined eighty-nine nations, from Austria to Zimbabwe, in banning use of the steel-jaw leghold trap. Internationally, Austria has banned fur ranches, both Denmark and Norway have declared that ranch-raised fur is "ethically unacceptable," and the British government has declared its intention to pass legislation to prohibit all fur farming. In the United States House of Representatives, legislation that would ban the use of the steel-jaw leghold trap on all federal lands garnered eighty-nine cosponsors from both major political parties. All the indicators point to the fur industry's steady downward spiral. Fur, once as "in" as anything could be in the world of fashion, increasingly is "out."

American consumption of most varieties of meat also is declining. Whereas fourteen million veal calves were slaughtered in 1945, the number declined to eight hundred thousand in 1995. Except for poultry, over-all per capita meal consumption continues to decline. USDA figures for "red meat" (beef, lamb, veal, and pork) for 1996 and 1998 were 119.5 and 112.0 pounds, respectively; fish, 15.1 and 14.7 pounds; and poultry, 51.9 and 64.3 pounds. This same period has witnessed a decline in per capita consumption of eggs and dairy products. Granted, some people who have stopped eating meat and meat products, or who have decreased the amount that they eat, have done so for reasons other than respect for animal rights. Legitimate health and environmental concerns, for example, can lead some people to make changes in their diets. Nevertheless, the national trend away from an animal-based diet and toward one richer in vegetables, legumes, grains, and nuts is unmistakable.

Is reliance on the animal model in research, testing, and education undergoing a comparable transformation? Because the numbers are hard to come by . . . no one can say with certainty. What is known is that the research community is increasingly willing to look for ways of replacing animals in the lab, researchers experience accelerated success in finding them, and a steadily rising number of Americans want to see this happen. A 1996 poll conducted by the Associated Press and the *Los Angeles Times* found that 72 percent of those responding said that it is sometimes wrong to use animals in research, and fully 29 percent said it is always wrong.

Even the American public's attitude toward the idea of animal rights is changing. Once the object of ridicule and sarcasm, animal rights increasingly is accepted as an appropriate moral norm. According to the poll just alluded to, fully two-thirds of adult Americans agree that "an animal's right to live free from suffering should be just as important as a person's." Even the courts are beginning to respond. In the past, advocates of animal rights have been prevented from having their case heard because they have not had legal standing. A recent verdict of the U.S. Court of Appeals for the D.C. Circuit reversed this pattern of denial. Henceforth, individuals and advocacy groups will have the legal freedom to bring suit against the USDA on grounds of its failure to enforce the provisions, limited though they may be, of the Animal Welfare Act. And two determined English activists, Dave Morris and Helen Steel, the "McLibel two," in the longest (314-day) trial in English history, successfully defended themselves against charges of libel brought by McDonald's. Among the court's findings: McDonald's is "culpably responsible for animal cruelty."

Is it, then, hopelessly unrealistic to imagine a day when fur coats will follow whale-bone corsets into fashion oblivion, when slaughterhouses will exist only in history books, and when all the scientific laboratories of the world will have a sign over their entrance proclaiming "No Animals Allowed"? Those who are pessimistic about the moral possibilities of humanity will answer yes. And perhaps they are right. But those who believe in the human capacity to change one's whole way of life, because both justice and compassion require it, will answer, no. Not in my lifetime, perhaps, but someday surely, I believe, the principled journey to abolition will be complete. As the evidence presented in the previous paragraphs suggests, for many people who understand and respect other animals, that long journey already has begun.

# 2

# Animals Are Not Entitled to Rights

## David R. Carlin

*David R. Carlin is a professor of philosophy and sociology at the Commu-
nity College of Rhode Island and a former state senator of Rhode Island.*

Animals currently have no legal rights. Animal rights activists,
however, believe that only a small biological difference exists between humans and animals. For example, many activists argue
that both humans and animals, such as dogs, demonstrate rational thinking. Charles Darwin's theories about evolution have
been used to reduce humanity to biological terms only. Specifically, Darwin's followers have deduced that because humans and
animals descended from the same ancestors, and because humans
strongly resemble nonhuman primates, there is little difference
between animals and humans. From a Christian perspective, this
reductionist view attacks "an ancient and exalted conception of
human nature" in which man is made in the image of God. For
that reason, the motives of those who seek animal rights are anti-
Christian. To consider providing animals with rights would be a
perilous road for humanity.

Harvard Law School [in 1999,] offered its first-ever course on animal
rights. This is good news for animal rights advocates, since Harvard
is one of the two or three top law schools in the nation. If Harvard is on
board for animal rights, can the Supreme Court be far behind?

Currently, American law gives animals protection in a wide variety of
circumstances, but it affords them no rights. The prevailing legal principle is that only persons can be bearers of rights. So, before animals can
have rights, either that principle will have to be changed, or it will have
to be shown that animals (at least some of them) are persons.

The animal rights movement (of which Peter Singer, the controversial Princeton professor, is the philosophical guru) contends that there
should be only a relatively narrow legal gap between humans and animals. Biologically speaking, of course, there is only a narrow gap between

David R. Carlin, "Rights, Animal and Human," *First Things: A Monthly Journal of Religion and Public
Life*, August 2000, pp. 16–19. Copyright © 2000 by the Institute on Religion and Public Life.
Reproduced by permission.

humans and the highest of the animals. But this raises the question: Is a strictly biological account of human nature adequate? The animal rights movement would answer this question in the affirmative; Christianity, by contrast, has always answered it in the negative. At first glance, the animal rights movement seems to be aiming at the elevation of animals. In fact, however, it is but the latest episode in a long history of attempts to degrade humans.

Many individual members of the animal rights movement, I willingly concede, are kindhearted folks who are revolted at cruelty to animals and wish to minimize it; they have no desire to degrade humanity. But historical movements often have objective tendencies that contradict the wishes of their proponents. (Witness communism, which, despite its objective tendency to tyranny and mass murder, had many followers who were humane and philanthropic in intention.) Underlying the push for narrowing the legal gap between humans and animals is the philosophical premise that there is no more than a narrow ontological gap between humans and animals. But the animal rights people are not the first to embrace this premise. Far from it.

## A history of animal rights philosophies

In the sixteenth century, Michel de Montaigne, the great French essayist and skeptic, argued that the gap between humans and animals was narrower than most people imagined. He devoted much of his writing to showing that humans are not nearly as rational as we, in our pride, suppose ourselves to be, while occasionally pointing out how surprisingly rational the lower animals could sometimes be. In his most comprehensive and influential essay, "An Apology for Raimond Sebond," Montaigne cited the case of a logical dog, a case reported by an ancient philosopher. The dog was following a scent along a path. Suddenly the single path divided into three. The dog hesitated: Which way to go? He sniffed at one path; no scent. He sniffed at a second; no scent there either. And then, without bothering to give an investigatory sniff at the one remaining, he set off on this third path. Clearly the dog had performed a disjunctive syllogism [deductive reasoning], saying to himself: "The scent I'm following will be found either on path A, B, or C; it is not found on A or B; it follows, therefore, that it must be on C."

> *The animal rights movement . . . is but the latest episode in a long history of attempts to degrade humans.*

And since, according to the dominant philosophical tradition of Montaigne's day—a tradition that reached back to Plato, Aristotle, and the Stoics—rationality (or a capacity for logical thinking) is the distinctive characteristic of human beings, it was no small thing to show that dogs as well as humans can be logical. In the world of philosophy, it had always been rationality that established the almost infinite ontological gap between humans and animals. Show that rationality is a characteristic shared by both, and humanity's ancient claim to dominance is destroyed.

Near the middle of the eighteenth century, during the robust early stages of the Enlightenment, a minor French philosophe, Julien Offray de la Mettrie, wrote a book titled *L'Homme Machine* [*The Human Machine*]. If humans are nothing more than machines, he argued, albeit very refined and complex ones, then there is certainly no great ontological gap between humans and the lower animals, for they are also machines, though less refined and complex. La Mettrie suggested, for instance, that the reason apes cannot speak is not because of any inferiority in rationality to human beings but because of "some defect in the organs of speech." He believed a young ape could be taught the use of language if we were to instruct it using the (then newly invented) methods used to teach deaf-mutes to "speak." In other words, given the right teacher, apes could be taught sign language.

## Darwinism and animal rights

But to date, the greatest of all attempts to narrow the gap between humans and the lower animals has been Darwinism. Perhaps this should not be said of the Darwinism of [naturalist Charles] Darwin himself, who had little wish, at least in public, to extrapolate his biological findings into the realm of ontology. But it can certainly be said of many of Darwin's epigones, who viewed humans as purely biological entities and thus regarded biology as competent to pronounce the last word on the ontological rank of human nature. Since humans have the same remote ancestry as the rest of the animal kingdom, since we have the same relatively proximate ancestry as the great apes, and since anatomically we bear a strong resemblance to these our "cousins," then it follows (they reasoned) that humans are ontologically only a little bit superior to the lower animals. And if we measure superiority and inferiority in terms of capacity to survive (which is perhaps the true Darwinian way of measuring these things), then we are not superior at all; for it is obvious that all surviving animal species have equally met that test. By that measure, our superiority, if we are indeed superior, will not be shown until we outlast all other animal species; but that is almost certainly impossible, since it is difficult to imagine how humans could survive on earth without the assistance of other simultaneously existing animal species.

Our contemporary animal rights movement is heir to this long tradition of trying to narrow the gap between humans and lower animals. But what motive lies behind this tradition? The answer seems obvious enough. Specifically, the motive is anti-Christian; more generally, it is a strong animosity toward the view of human nature taken both by biblical religions and by the great classical schools of philosophy, especially Platonism and Stoicism. That man is "made in the image and likeness of God" is an expression found in the Bible, but it is a formula that well expresses the anthropology of Plato and the Stoics as well. To reduce human nature to nothing more than its biological status is to attack this ancient and exalted conception of human nature.

In defense of the attackers—from Montaigne, through the philosophes and the Darwinians, to Peter Singer (who once wrote a book titled *Animal Liberation*) and the Harvard Law School—it might be said that their intentions have often been humane. The Stoic-Christian theory of human nature, in their opinion, has been dangerously unrealistic, the product not of empirical observation but of fantastic imagination. By encouraging men and women to believe that their true home is not in this world, the world of nature—that we are potentially divine beings living in temporary exile—this fantastic theory has rendered humans unable to achieve such limited happiness as we might have achieved. Demoting human nature from heaven to earth will, by making us more realistic, render us more successful. Better (in [English novelist Dame Rose] Macaulay's phrase) to own an acre in Middlesex [England] than a county in Utopia.

This defense ("they had good intentions") might have been acceptable prior to the twentieth century. But in the course of that century we had some unpleasant experiences with persons who entertained the purely biological conception of human nature. Hitler was a great believer in this purely biological conception (sometimes with a confused overlay of pagan romanticism). In his way, he can be counted as one of Darwin's epigones. Now, of course, you cannot prove that an idea is wrong simply because Hitler embraced it; for instance, that Hitler favored the production of Volkswagens doesn't prove that they are bad automobiles. But when there is a direct link between one of his major ideas and the Holocaust, as there is in the case of his conception of human nature, this is at least enough to give us pause. At present I cannot prove that the idea of animal rights is extraordinarily dangerous and inhumane; to get proof of this, we'll have to wait until the disastrous consequences of the idea reveal themselves over the next century or so. But I strongly suspect that it's a dangerous idea, and accordingly I suspect that the promoters of this idea, whatever their intentions, are enemies of the human race.

# 3

# Animal Testing Is Essential for Medical Research

## Lawrence Corey

*Lawrence Corey is professor of laboratory medicine at the University of Washington School of Medicine in Seattle, Washington, and head of the infectious diseases program at the Fred Hutchinson Cancer Research Center.*

Animal testing is essential to drug and vaccine research. In particular, animal experiments have been vital in discovering drugs that slow the progress of the human immunodeficiency virus (HIV), the virus that causes AIDS. Similar advancements have occurred in developing treatments for herpes and hepatitis B because of animal testing. New methods of research such as computer modeling and in vitro testing have helped reduce the use of animals in biomedical research in the last twenty years, but animal experimentation is still needed to prevent harm to humans from new medicines or vaccines. Without animal experimentation, human lives would be jeopardized.

The past 20 years have brought remarkable progress in the development of therapies and vaccines for treating viruses.

When I began doing research, there were only two anti-viral medications available, and both were rarely used. [In 2000,] there [were] 14 licensed anti-viral drugs for treatment of human immunodeficiency virus (HIV) infection alone.

One needs only to look at a picture of [basketball legend] Magic Johnson—who has tested positive for the AIDS virus—during a visit to a central Seattle Starbucks to appreciate what these drugs have done to help people.

In the United States, Acquired Immune Deficiency Syndrome, or AIDS, which is the disease caused by HIV infection, has gone from a rapidly fatal disease to one that can be slowed significantly by drug therapy. Likewise, anti-viral drugs for herpes virus infections have reduced suffering from lesions caused by herpes simplex, and they have markedly reduced death from viral pneumonia in transplant patients and viral-related transplant rejection.

Similarly, we now have a vaccine against cancer. Hepatitis B vaccine prevents hepatitis B and liver cancer, its major complication.

How has this progress been possible, and what role do research animals and alternative forms of research play in this progress? Do we need animals at all?

## The changes in animal experimentation

In many ways the latter half of the 1900s can be described as the time of development and widespread use of animals in research. Mice were the mainstay of this type of research, to help us understand what caused the cancer and how to stop it. The past 20 years have seen a reduction in the number of animals used and the development and use of alternatives, including elegant cell culture models and computer models.

There have been some real changes. The cancer-causing potential of drugs is now tested first in bacteria developed for this purpose and only then in mice or rats. And nearly all testing of cosmetics in animals has stopped. But we still do use animals. Why?

The popular press would have us think that medical breakthroughs come from giant "insightful leaps." In fact, dramatic improvements in medical therapy are made in small, incremental steps by large teams of scientists. But this process is not short. Nor is it smooth or predictable. For all the novel therapies I am aware of, experiments of a candidate antiviral or vaccine showed a glimmer of an effect in the test tube, but not enough to move to the next stage.

Those compounds that work in a test tube—in vitro—must then pass the test of activity and tolerance—called toxicity testing—in a whole animal. More than 90 percent of compounds that have activity in the test tube against infection or tumor cells flunk animal toxicity studies. There are no substitutes for testing in animals to measure the potential harm of new drugs. These experiments help us see how risky a compound is for use in people, and at what dose.

You may recall the gene therapy case of Jessie, who died of organ failure when an experimental virus was used, and we later learned that animal tolerance to this virus was low. This tragic case reminds us that when a treatment does cause poisonous effects in animals, extreme caution is warranted. Animal studies are even more crucial for developing medicines for young infants and children because their rapidly growing cells make them more susceptible to some toxic drugs.

What about computers? Biochemists have developed computer models to look at relationships between drugs and their targets in an effort to build better *keys* to fit the molecular *locks*. However, predictions based on those models are imperfect at best.

## A close look at two medications

Let's look at [an] example of two drugs I prescribe every day for my patients: acyclovir and ganciclovir. I remember the first time I used the acyclovir medication on an infant with neonatal herpes. The previous infants I had seen had died. This one miraculously started to get better four days into the treatment, something I had never seen before.

Now, acyclovir is the most-used anti-viral drug in the world—a very effective treatment for genital herpes and neonatal herpes. Hundreds of my patients take it daily.

Ganciclovir was discovered two years after acyclovir and, in the test tube, looked like a better compound. At the molecular level, the compounds were almost identical. Yet in animals the two were very different. Ganciclovir killed bone marrow cells; acyclovir did not. Ganciclovir caused sterility in animals; acyclovir did not. Human results matched the animal tests.

Acyclovir is one of the safest drugs we have in my field; people can take it daily for years. Ganciclovir has a role in treating transplant and HIV-infected people. It is a life-saving and eyesight-saving drug, yet its strong toxicities limit its use to those with severe illnesses. All this was defined by prudent animal testing.

## The use of animals in vaccine testing

How about the role of animals in vaccine testing? Vaccines protect people from disease by stopping infections before they can wreak their havoc. Vaccines ultimately are tested directly in people, so why not skip animals?

Let's take as an example the development of a vaccine against HIV to prevent AIDS, which is devastating the African and Asian continents with 16,000 new cases a day and continues to spread throughout the world unchecked.

Because vaccines work by stimulating the body's immune system to fight back against the virus, there is no way to test in cells, or on a computer, how the vaccine will work in the whole animal. New vaccines are given to experimental animals, followed by a "challenge" with the infectious agent one wants to prevent. These animal model experiments define whether novel vaccines are safe enough to initiate clinical trials.

More important, they show whether the vaccine is good enough to justify large-scale testing, which may involve tens of thousands of people and cost tens of millions of dollars. Non-human primates, especially macaque monkeys, are critical for the development of an HIV vaccine.

The work to translate results from the monkey models to vaccines that can go into humans involves intense communication between those of us involved in human vaccine development and those laboratory researchers involved in developing candidate vaccines.

The studies do not substitute for vaccine testing in humans. Without the tests, and without this dialogue, many entirely ineffective vaccines might be tested in people without any benefit, wasting money and time. Does it not seem wise, then, to know what a vaccine does in a primate challenge study before we administer it to thousands of people?

We can and do use alternatives at each step in our process of drug and vaccine discovery and testing, to refine our choices before we go into animals and people. The reality of developing novel therapies and vaccines for human disease is that prudent use of animal resources is a necessary part of the process of medical research to improve human and animal health.

# 4

# Animal Testing Is Not Essential for Medical Research

## C. Ray Greek and Jean Swingle Greek

*C. Ray Greek and Jean Swingle Greek are cofounders of Americans for Medical Advancement, a nonprofit organization, which educates the public about the hazards of applying the results of animal testing to humans. They are also authors of the book* Specious Science: How Genetics and Evolution Reveal Why Medical Research on Animals Harms Humans.

Modern medical advances such as antibiotics and vaccines are not the result of animal experiments. For example, experiments with mice and rats failed to turn up any connections between cancer and smoking. Epidemiological studies, not animal experiments, found links between heart disease and cholesterol. Furthermore, more than half of the medications released between 1976 and 1985 were taken off the market or relabeled because dangerous side effects were discovered that had not been found in animal experiments. AIDS research with primates has also shown a high level of failure. Instead of relying on animal experiments for their research findings, scientists should use other, more dependable, techniques such as in vitro testing, modeling studies, and clinical research. Animal experiments continue only because they are profitable.

Medical advances are responsible for Americans living longer and better lives. But where have these modern-day miracles come from? Those who profit from animal experimentation, like at Colorado University [CU], would have us believe they came about as a result of research conducted on animals. Franki Trull, a representative for the animal research industry, has even stated that, "every major medical advance of this century has depended on animal research." But what are the facts? Animal experiments were not responsible for vaccines, MRI [magnetic resonance imaging] and CAT [computer-aided tomography] scanners, anesthesiology, antibiotics, medications that combat AIDS, chemotherapy, or modern surgical techniques. The lack of scientific support for extrapolating the results of animal experiments to humans speaks for itself.

---

*The medications used to treat heart disease and high blood pressure were developed despite misleading results of animal experiments.*

---

Experiments on animals did not link heart disease to cholesterol, or high blood pressure to strokes. Epidemiology did. The medications used to treat heart disease and high blood pressure were developed despite misleading results of animal experiments.

Experiments on animals did not find any link between cancer and smoking. Cancer research has an abysmal record of failures when using rats, mice, and other animals. Of 20 compounds known not to cause cancer in humans, 19 did cause cancer in animals. On the other hand, of 19 compounds known to cause oral cancer in humans only 7 caused cancer in mice and rats.

Animal experimentation did no better in the field of surgery. Radial keratotomy is a surgery performed to enable better vision without glasses. The first radial keratotomies were animal experimentation-induced catastrophies. Surgeons thought they had perfected the procedure on rabbits, but it blinded the first humans.

Medication testing is another oft-cited example of animal necessity in medical science. But consider this: Of the 198 new medications released between 1976 and 1985, 102 were either withdrawn or relabeled secondary to severe side effects not predicted from experiments on rats, mice and other animals. These side effects included complications like lethal dysrhythmias, heart attacks, kidney failure, seizures, respiratory arrest, liver failure, stroke and many more.

But let's discuss some animal experiments taking place at CU. Researchers are using monkeys to supposedly conduct research on AIDS. They have spent 20 years and nearly seven million taxpayer dollars to take baby monkeys away from their mothers. That's right, they think that by depriving infant monkeys of their mothers they will somehow find a cure for AIDS. How ludicrous. True, the immune system is adversely effected by stress. We have known that for years based on clinical observation of humans. But to think that by studying maternal deprivation in monkeys one can find the cure for AIDS is like thinking that by studying the wheel you can propel man to the moon.

Experiments on primates have a long history of misleading scientists about AIDS. Primate experiments misled researchers about how rapidly HIV [human immunodeficiency virus, which causes AIDS] replicates resulting in mistreatment and lost lives. The current medications used to treat AIDS were discovered in a test tube and bypassed animal testing altogether.

People often ask, "what should we do if not experiments on animals?" We hope the answer is obvious. We need more funding to do more of the research that got us here in the first place. We now enjoy the highest standard of medical care secondary to epidemiological studies, in vitro research, clinical research and observation, autopsies, research conducted on human tissue, mathematical modeling studies, and technological advances. These are the only techniques that have worked in the past and it is what we should be funding now!

People who have a vested interest in animal experimentation will simply dismiss the facts in this [viewpoint] as being ridiculous. They will present their version of the truth in such a way as to make it appear that animals have been invaluable to the scientific process. If anyone were willing to present their opinion in an open forum, we would gladly agree to debate them. The truth frequently comes out when views must be explained openly in public.

Animal experimentation does not continue because of the great medical strides that are falsely attributed to it. The practice continues for one reason. People make money from doing it. Animal experimentation is a multi-billion dollar industry. Every time a researcher receives money for experiments on animals, the university where he is employed takes a percent off the top. This money can then be used by the university in virtually any way it wants.

Unless someone from CU accepts our challenge to a public debate, the issue must be decided based on the following: Whom would you rather believe? The medical historians, former animal experimenters, human researchers, scientists, physicians and veterinarians who have stated that animal research is futile or the people and big businesses making billions of dollars every year from experimenting on animals?

# 5

# Animal Testing Is Cruel and Does Not Benefit Medical Research

## Ingrid Newkirk

*Ingrid Newkirk is cofounder and president of People for the Ethical Treatment of Animals (PETA).*

Millions of animals suffer through stressful and unnecessary tests every year. Numerous examples indicate that these experiments are wasteful, cruel, and ridiculous. In one study, pregnant rabbits were given cocaine and their offspring received shocks in order to study maternal drug use. In another study, cats were shot in the head to show that this type of wound impairs breathing. The nature of some experiments has not changed in over seventy years. Most researchers defend their actions and the actions of their colleagues, although a few have resigned in protest over the horrendous treatment of their test subjects.

Some months after founding People for the Ethical Treatment of Animals (PETA), Alex Pacheco led the police into a laboratory called the Institute for Behavioral Research (IBR), in Silver Spring, Maryland. Once inside, police officers served their warrant, seizing seventeen small macaque monkeys, survivors from a group originally twice that size.

The monkeys were in bad shape. Many had open, festering wounds, and much of their lustrous hair was missing. Their normally bushy tails were bare because of malnutrition, and they had pulled out whole clumps of fur on their arms and legs from frustration, anger, and misery. Although once vigorous and even fierce defenders of their jungle homes, after years of confinement in feces-encrusted cages barely larger than their own bodies, they were now frail and vulnerable. They stared up anxiously at the crowds of uniformed officers and media, almost blinded by the sunlight they had not seen since being snatched, many years before, from their families in the Philippines.

As the state's veterinary witnesses would later testify in court, many of the monkeys had been operated on, their backs cut open and their nerves severed, making movement of their arms difficult or impossible.

One timid little monkey named Billy had not only lost the use of both his arms, but both were broken. He had been forced to push himself on his elbows across the cage grating and to eat his food by bending over and grasping it between his teeth, although his teeth were painfully infected.

Alex would relate how the monkeys would injure their deadened limbs, sometimes catching and then tearing off their fingers on the jagged, broken, and rusted wires that protruded from their cages. (Police documented thirty-nine of the fingers on the monkeys' hands were severely deformed or missing.) He recounted how the experimenters forced the monkeys into a dark, blood-spattered refrigerator and a jerry-rigged restraint chair, tying them down with duct tape and burning them with a cigarette lighter, squeezing their flesh, including their testicles, with surgical pliers, and administering electric shocks to them to "test" the feeling in their limbs.

Was the Silver Spring case an isolated incident?

Most people believe, or want to believe, that animals would not be used in experiments if their use weren't absolutely necessary. After all, who wants to believe we are needlessly cruel? Moreover, people hope that only a few animals, or as few as possible, are used, that all of them are destined for experiments that are potentially life-saving, and that they are treated humanely.

Nothing could be further from the truth.

## The extent of the suffering

Millions of animals are used in experiments every year. More often than not, they are acquired almost casually, housed abysmally, and denied anything remotely like a life. In addition to suffering through the experiments, they are under constant stress from fear, the loss of control over their lives, and the denial of all that is natural and meaningful to them, such as enjoying the company of others of their own kind and choosing.

*Animals from giraffes to gerbils are used for everything from forced aggression and induced fear experiments to tests on new football helmets and septic tank cleaner.*

Animals from giraffes to gerbils are used for everything from forced aggression and induced fear experiments to tests on new football helmets and septic tank cleaner. Baboons are given AIDS-infected rectal swabs, great apes are purposely driven mad to make them crush their infants' skulls in child abuse studies, and researchers are changing the genes of pigs so they can no longer walk and chickens so they can no longer fly.

Animals are burned alive in the cockpits of planes, exploded in weapons tests, and forced to inhale pollutants until they choke to death. They are starved and shot; they have hallucinogenics and electrical shocks administered to them; they are force-fed poisons and used to demonstrate already well established surgical procedures. They are commonly thought of as nothing more than disposable "test tubes with whiskers."

There are countless examples of wasteful and ludicrous experiments. This "rubbish research" comes at a time when many Americans do not have health insurance, scores of alcohol and drug treatment clinics have closed due to the loss of funds, the elderly and disabled go without new eyeglasses or dental care, and unless they can afford to buy them themselves, disabled people are left without state-of-the-art wheelchairs and home aids that would allow them to participate more fully in society.

## Ridiculous studies

At the University of California in Santa Barbara, experimenters sewed plastic swords to the hindquarters of male fish to see if females preferred males with or without swords (try to guess the applicability of this jewel). Elsewhere, rats were killed by being fed huge doses of Louisiana hot sauce (the human equivalent of half a cup per 10 pounds of body weight). Monkeys at the University of Texas had electrical probes inserted into their brains and were then awakened every night with loud noises to see the effect on their sex drive; rats were forced to swim to their deaths at Georgetown University to study "executive stress" (female rats stay afloat longer); more University of California studies had experimenters shoving toy snakes into monkeys' cages to see how the monkeys reacted (they were scared); and fish were given a choice between gin and vodka (whatever their preference, who cares?).

Forgetting the just plain daft experiments, there are tens of thousands of experiments that cause enormous suffering to animals every day in the name of medical research. For example,

• at Rockefeller University, experimenters have forced cats to vomit up to ninety-seven times in three and a half hours after severing the connections between the cats' brains and spinal cords;

• at the University of Iowa, pregnant rabbits have been given daily doses of cocaine, and baby rabbits were shocked in the head to study "maternal drug abuse";

• at Louisiana State University hundreds of cats were shot in the brain to show that such wounds "impair breathing";

• the U.S. Fish and Wildlife Service got into the act by spending $600,000 after the Exxon Valdez oil spill to capture birds, shoot them, outfit them with radio tracers, douse their corpses with oil, and throw them into the sea to "prove" that birds were killed by the spill;

• University of Illinois researchers cut open cows' stomachs, inserted bags of newspapers into them, then checked the bags to see if cows can survive on a diet of 40 percent newsprint;

• NASA has sent monkeys into space with electrical coils threaded through the backs of their eyes;

• the tobacco industry has forced dogs and mice into smoking masks and compelled them to inhale tobacco fumes twenty-four hours a day for years;

• half a dozen universities have kept cats awake for days at a time, forcing them to balance on narrow planks above water-filled tanks or lowering their cage temperatures to well below freezing.

## The "standard four" tests

As if that weren't enough, hundreds of household product and cosmetics companies and big pharmaceutical houses, many marketing the fortieth version of basically the same old antidepressant or headache remedy, still contract with laboratories that conduct the "standard four" tests. These crude tests, first hurriedly devised between 1920 and 1930, when large quantities of new pills and locations were flooding the market, are carried out on large groups of animals every day.

Here's how unsophisticated these tests are: You take a substance—say, an acne medication or a nail polish remover—and (1) drip it into restrained rabbits' eyes; (2) thrust massive quantities of it down dogs' or monkeys' throats; (3) force rats to inhale it through a mask or by spraying it into their sealed cages, and/or (4) smear it on the raw, shaved backs of guinea pigs. Then you sit back and record the damage.

*There are tens of thousands of experiments that cause enormous suffering to animals every day in the name of medical research.*

Obviously, if you force-feed a pint of drain cleaner to a monkey, the monkey's internal organs will be eaten away, along with the lining of his or her throat, and he or she will probably go into convulsions and die. His or her pain will not soothe or save the person who tries to commit suicide by swallowing drain cleaner, nor will it help in the least little bit the person who, somehow or other, inadvertently gets drain cleaner in his or her eye. We already know the likely result and the proper treatment, and these tests are not designed to determine how to treat injury.

## One case of laboratory abuse

Rabbits, guinea pigs, and other small animals suffer silently, their convulsions violently wracking their bodies without disturbing the peace of the rooms in which they "live": rooms containing row after row of plastic "shoeboxes" or stainless steel cages. The dogs will have been debarked—their vocal cords severed to cut down on the noise.

Add to all this the sort of scenes filmed secretly in 1998 inside the British headquarters of Huntingdon Life Sciences, a laboratory that tests for some of the top names in the industry worldwide. Workers were caught punching dogs in the face, screaming at the animals, and even simulating sex with each other while trying to inject a frightened beagle. In the United States, Huntingdon moved quickly in court to prevent PETA from showing exactly what its investigator had videotaped when she worked in their New Jersey facility the same year. The bubble of public assurances that animals are well cared for was in danger of being burst.

Luckily, the PETA investigator's tapes were aired on television in Cincinnati, Ohio (home of Procter & Gamble), and Norfolk, Virginia (home of PETA), before the gag order took effect. The tapes showed monkeys who were scared out of their wits being taped to an operating table.

Their fear is almost palpable. The animals were supposed to be kept calm and quiet to facilitate accurate readings from an electrocardiograph, yet joking staff played loud rock music, yelled in the faces of the restrained monkeys, yee-hawed at the top of their lungs like drunken cowboys, and stuck lotion bottles into the helpless animals' mouths. Following these high jinks, workers body-slammed the primates back into their steel cages.

Should you harbor the illusion that most researchers would immediately denounce such cruelty, dream on. They will defend almost anything, as indeed the top U.S. research lobbying groups did, openly applauding Huntingdon for trying to bar PETA from allowing the public to see the photos and tapes and judge for itself. In fact, Huntingdon's injunction went so far as to specifically prohibit PETA from answering the federal government's questions about the lab or giving the tapes to members of Congress. However, Huntingdon was too late: PETA had already turned over its eight months of investigation notes, hundreds of photographs, and hours of videotapes and had filed a thirty-seven-page formal complaint with the U.S. Department of Agriculture (USDA).

Federal authorities found Huntingdon in violation of the Animal Welfare Act (AWA), the only federal law that offers any protection whatsoever to animals in laboratories.

## Researcher responses

Perhaps some researchers defend abhorrent practices because they are inured to the suffering of those around them. Perhaps, just as some people can stare at a Picasso for hours without "getting it," they can't understand for the life of them what all the fuss over animals is about. Or perhaps they see any acknowledgement that there are problems as a chink in the armor that could one day crack the whole suit.

There are certainly some truly arrogant experimenters, like Robert White, the Cincinnati man who performs horrific head transplant experiments on monkeys and watches the anguished eyes on their disembodied but live heads follow him about the room. One morning in Cincinnati, White told me it is "never acceptable" to put limits on science. Later, when he and I addressed a group of high school students, White railed to them, "Who are *you* to question a scientist?"

The Silver Spring monkeys prosecutor, Roger Galvin, learned how researchers unite to protect their own kind, like a mob defending itself against the authorities. When given every opportunity to distance themselves from the disgusting cruelty, filth, suffering, and unscientific conduct in that case, what did the research community do? Did it condemn a facility in which police found dead monkeys floating in barrels of formaldehyde, their infected limbs and rotted bandages weighed down with auto parts? Hardly. Its distinguished members flocked to the courthouse to testify *in support of* the accused, Dr. Edward Taub, a researcher many of them had never met or spoken to.

George Bernard Shaw may have hit the nail on the head when he said, "He who would not hesitate to vivisect, would not hesitate to lie about it." After each glowing endorsement, Mr. Galvin would ask the witness for the defense whether he or she had bothered to visit this laboratory. "No." "Did you ask to see the police photographs?" "No." "Did you bother to speak to the state attorney's office about the charges?" "No." The defense witnesses mocked concerns for the animals, describing the monkeys as "nothing more than defecating machines" and a cockroach infestation as an "ambient source of protein" for the primates. They described Taub as a "modern day Galileo," an apt comparison perhaps, given that, in his time, Galileo's opinions allowed experimenters, like the evil René Descartes, to nail live animals to a board, eviscerate them, and disregard their screams as nothing more important than the sound of squeaking wheels.

As with everything, there are exceptions. One is Donald Barnes. Once an experimenter for the United States Air Force, Barnes used to torture rhesus monkeys. His job was to irradiate them, then strap them to treadmills to run, vomiting, to their deaths. Over time, a light came on in his head. He realized that the experiments had yielded nothing of use and were simply a line item on the base budget that his superiors would never sacrifice.

Barnes also came to see that what he was doing was unethical. In his own words, he decided one day to "call in well," and never went back. Mr. Barnes testified for the prosecution in the Silver Spring monkeys case.

Dr. Roger Ulrich is another example. Dr. Ulrich received many professional awards and honors for his rather nasty research, using monkeys to study the relationship between pain and aggression. One day, he wrote to the American Psychological Association.

> When I was asked why I conducted these experiments, I used to say it was because I wanted to help society solve its problems of mental illness, crime, retardation, drug abuse, child abuse, unemployment, marital unhappiness, alcoholism, over-smoking, over-eating . . . even war! Although, after I got into this line of work, I discovered that the results of my work did not seem to justify its continuance. I began to wonder if perhaps financial rewards, professional prestige, the opportunity to travel, etc., were the maintaining factors and if we of the scientific community, supported by our bureaucratic and legislative systems, were actually part of the whole problem.
>
> One spring I was asked by a colleague, "Dr. Ulrich, what is the most innovative thing that you've done professionally over the past year?" I replied, "Dear Dave, I've finally stopped torturing animals."

# 6

# Animal Testing Is Becoming More Humane

## Erik Stokstad

*Erik Stokstad is the managing editor of* ScienceNOW, Science *magazine's online news service.*

Many new technologies are being developed that are making animal testing more humane and reliable. Historically, scientists would anesthetize animals in order to provide test drugs orally and then monitor the animal's bodily functions. Sometimes scientists would infect mice with a disease, give them antibiotics, and then kill two mice every two hours to evaluate the medicine's effects. Now new sensors and monitors can be implanted in an animal to transmit data. Because these new sensors are less stressful for the animal, the data produced is more accurate and fewer animals are killed. New imaging techniques also reduce the animal's distress because they are less invasive than surgery. Imaging allows scientists to track disease in an animal by scanning for infections or tumors. Human stem cells and DNA "chips" are also being studied for testing because they are more reliable and versatile than earlier techniques. Unfortunately, because regulatory agencies are slow to validate new testing, it is difficult to persuade scientists to use the new procedures and technologies.

For decades, more and more researchers have been using fewer laboratory animals for compassion's sake. Thanks to new experimental techniques, many are getting cleaner results, too.

[In 1989], veterinary surgeon Christian Schnell tested candidate drugs to lower blood pressure with a procedure that was highly stressful—for himself and his test animals. First, he would anesthetize marmosets and insert a catheter into an artery in their legs. The next day he restrained the conscious animals, orally administered the drug, and recorded blood pressure through the catheter for 4 to 5 hours. Not only did the harried animals' hearts race during the experiment, but each one could be used for only six trials before all its suitable arteries had been tapped.

But by 1991, Schnell, a researcher at the drug company Ciba-Geigy in Basel, Switzerland, switched to a new and more sophisticated technology: a sensor that he implanted in the animals' abdominal cavities. The device continually measures blood pressure and transmits the data to receivers in the cages, allowing the marmosets to move freely and remain with their families—more relaxed and with normal heart rates. Without the confounding effect of stress, the results are cleaner. "We are now convinced we're measuring the truth," says Schnell. And without the need for catheters, Schnell could do the same research with only 10% of the marmosets he had previously needed, saving the company up to $200,000 a year.

## A new technological trend

Schnell's case illustrates an accelerating trend in which new technology is helping researchers reduce their reliance on animal experiments, while at the same time improving their results. Although animal rights extremists continue to use violent and intimidatory tactics against researchers in many countries, more moderate campaigners for animal welfare have for years been working with researchers to encourage this trend toward better experimental design and more humane techniques. The motto of this movement is "Humane science is better science," and its creed is the "three R's"—replacing laboratory animals, reducing their numbers, and refining techniques to minimize pain and suffering. The results have been striking: The use of lab animals has declined in many European countries—in some cases by as much as 50% over the past 2 decades.

*New technology is helping researchers reduce their reliance on animal experiments, while at the same time improving their results.*

As a result, the mood among the more than 800 researchers who gathered [in Bologna, Italy, in August 1999] for the Third World Congress on Alternatives and Animal Use in the Life Sciences was cautiously upbeat. They exchanged information on a variety of technologies—including implantable sensors like those Schnell uses and new imaging techniques to replace invasive procedures—that are already reducing the number of animals and lessening distress. And researchers reported progress in several areas—such as DNA arrays and tests using stem cells—that could help drug companies rule out dangerous compounds before they're tested in animals. "The spin-offs of molecular biology and biotechnology will have a great impact on [lowering] the use of lab animals," predicts geneticist Bert van Zutphen of Utrecht University in the Netherlands.

## A few obstacles

But not all the trends are downward. Many animal welfare researchers are alarmed by the imminent prospect of a new round of toxicity tests in the United States on a host of so-called high production volume chemicals, as well as tests on potential endocrine disrupters, that may require millions of laboratory animals. And in some hot areas of research, such as transgenics, animal experimentation is rising fast. Since 1990, the number of procedures on transgenic animals in the United Kingdom, for example, has risen almost 10-fold to more than 447,000. That's "a huge rise and due to get much higher," predicts Caren Broadhead of the Fund for the Replacement of Animals in Medical Experiments in Nottingham.

Even when researchers come up with technologies that can lessen the use and suffering of test animals, they still face a formidable obstacle: the glacial pace of regulatory bodies in accepting replacement tests, such as cell cultures. "A validation study takes a long time," says Herman Koeter, principal administrator of the Environmental Health and Safety Division of the Organization for Economic Cooperation and Development in Paris. "You need years and years to get a gold standard." That frustrates researchers. "If people knew how onerous it can be to get a test validated, many fewer would begin developing new ones," says Ian Kimber, research manager of AstraZeneca's Central Toxicology Laboratory in Alderley Park, [United Kingdom].

## Advances for telemetry sensors

Schnell's work with marmosets to test potential blood pressure drugs is Exhibit A in support of the humane science movement's claim that compassion can improve science. Ciba-Geigy had been puzzled by the fact that some candidate compounds that had looked promising in the earlier, more invasive, tests were duds in early human trials. But when Schnell tried those compounds again using implanted monitors in unrestrained marmosets, they proved to be 10 times less effective at lowering blood pressure than they had in the restrained animals. "It was a shock when we discovered this," recalls Schnell.

Since those early tests, telemetry sensors have shrunk in size and price and they are becoming more widespread. Blood pressure monitors weighing 3.5 grams are now small enough to be implanted into mice, and the device that Schnell uses costs about $3000. The new monitors are also far more versatile: Implantable devices can record temperature, blood pressure, heart rate, electrocardiograms, and intraocular pressure, and blood flow monitors will be available soon. "I'm convinced that telemetry will be the standard method in the near future," says Schnell.

## Using imaging techniques

Whereas implantable monitors can keep track of an animal's physiology, an imaging technique developed by Xenogen Corp. of Alameda, California, allows researchers to chart the course of an infection or the growth of tumors without any surgery at all. The technique essentially records a glow from inside the animal. The light bulb is the luciferase gene, which produces the firefly's bioluminescent protein. Researchers infect an animal with a microbe engineered to express luciferase, anesthetize it, and place it in a dark chamber. Some of the photons from the luciferase pass through the animal's flesh, and a charged-coupled device counts them for a few minutes, pinpointing the active microbes.

The pharmaceutical industry is eyeing this technology as a potential replacement for a standard test called the mouse thigh model. To check out new antibiotics, for example, technicians give the test drug to 14 or more infected mice, then kill a pair of the animals every 2 hours, grind up their thigh muscles, and culture microbes from the tissue over 2 days. The better the antibiotic, the fewer microbes grow on the ground-up muscle. In contrast, researchers can scan a living mouse in just 5 minutes. And measuring the same animal throughout the study—rather than comparing individuals that might have had slightly varying initial infections or responses to the drug—also reduces variability.

---

*Not all the trends [in experimentation] are downward. . . . In some hot areas of research, such as transgenics, animal experimentation is rising fast.*

---

One group of researchers, led by Tom Parr of Lilly Research Laboratories in Indianapolis, compared the two techniques and presented their results at the 39th Interscience Conference on Antimicrobial Agents and Chemotherapy in San Francisco in September [1999]. The team ran a mouse thigh model using doses of a known antibiotic, but before extracting the muscle, they imaged the animals. The dose-response curves from the two assays were very similar, with correlations ranging from 0.94 to 0.98. Imaging "is more sensitive and more precise while requiring fewer animals," says Parr. "We should be able to get more valuable information in less time." The quick results also mean that test animals can be killed before they suffer the full effects of an infection. Xenogen president Pamela Reilly Contag says six pharmaceutical companies, including Eli Lilly, are evaluating the technology, and 10 others are in various stages of negotiation.

### Stem cell research

Drug companies are also showing interest in alternatives to animal tests to screen compounds for effects on fetal development. Researchers currently test for potential teratogenic effects [causing developmental malformations] by treating pregnant animals with a candidate drug and then checking embryos for abnormalities—a time-consuming and expensive proposition. "Most companies now want to have short tests that give a clear answer and that require small amounts of compound," says Philippe Vanparys, director of genetic and in vitro toxicology at Janssen Research Foundation in Beerse, Belgium. Recent developments in establishing immortal lines of stem cells—general-purpose embryonic cells that can develop into any type of cell in the body—have raised hopes that such tests may be feasible.

Because stem cells have a very reliable pattern of development into tissue, researchers can precisely measure any disruption to the number of cells, the quality of cells, and the timing of development. This provides a way of looking for subtle chemical effects that might lead to birth defects in particular organs. For example, Anna Wobus of the Institute of Plant

Genetics and Crop Plant Research in Gatersleben, Germany, has developed an in vitro method to differentiate mouse embryonic stem cells into heart muscle cells, among others. Once these cells begin to beat after 9 days of normal development, researchers can check for defects in the nascent heart. In 1996, Horst Spielmann, director of the National Centre for Documentation and Evaluation of Alternative Methods to Animal Experiments in Berlin, submitted this test to the European Centre for the Validation of Alternative Methods (ECVAM) in Ispra, Italy, an organization run by the European Union that assesses the suitability of in vitro tests for replacing established animal tests. "So far it looks very promising," says Juergen Hescheler, a molecular biologist at the University of Cologne, Germany.

---

*An imaging technique . . . allows researchers to chart the course of an infection or the growth of tumors without any surgery at all.*

---

Now, Hescheler and his colleagues have added a feature to the test that could make it even faster, easier to use, and more versatile. At the Bologna meeting, he reported that his group has spliced a fluorescent reporter gene[1] to the cardiac-specific promoter gene,[2] so the cells express a green fluorescent protein on day 4 of development, cutting experimental time in half. "We can directly measure cell differentiation without any staining, so it's less time-consuming," says Hescheler. The team now wants to link reporter genes to other types of stem cells, such as neuronal, epithelial, and cartilage precursor cells. If the reporter proteins could fluoresce in different colors, scientists might be able to examine the effects of potential toxicants on a suite of tissues at once. Interest in the cardiac reporter is already high. "In the last month, I had five to six pharmaceutical companies asking for this test," says Susanne Bremer of ECVAM.

## The new DNA chips

Toxicologists are also turning to a hot new genetics technology to study cellular responses to test compounds: DNA microarrays, which are commonly used to track patterns of gene expression. A single DNA "chip" carries an array of hundreds or thousands of short strands of DNA, each of which acts as a probe for a specific gene. To tell which genes were active in a sample, researchers convert messenger RNA to complementary DNA, tag it with a fluorescent marker, and wash the sample over the chip. The cDNA sticks to a specific probe on the chip, and its presence is revealed by a glowing patch when the chip is illuminated with light.

Many toxicologists believe that such arrays could reveal which genes a cell turns on in response to toxic compounds—and because they directly probe the activity of human cells, the arrays may eventually be better than animal tests in predicting toxicity to humans. "DNA chips will be the source of the next reduction in animals used," predicts Spielmann.

---

1. a nucleic acid sequence that encodes proteins    2. a part of a gene used to start the motion of DNA

AstraZeneca's Central Toxicology Laboratory (CTL) is one of the first off the blocks with a chip outfitted with DNA from 600 genes, associated with everything from cell adhesion and ion channels, to metabolism and immune response—all thought to be involved in cellular response to toxicity. "The most exciting thing about toxicogenomics is that we're going to start investigating genes we never would have thought of looking at," says CTL's Kimber. "That's where the big surprises—and big benefits—are going to come from."

Not everyone is convinced by the promise of DNA chips, however. "There's much hype about gene chip technology," says molecular biologist Johannes Doehmer of the Technical University of Munich in Germany. "They're very expensive, and it will take a few years before you can rely on them." And although the microarrays generate a lot of information very quickly, the results can be hard to interpret. "The vast majority of our time is [spent] figuring out the gene response" says CTL's William Pennie.

Even though many researchers say that animals will never be replaced for conducting general investigations or checking a whole-body response to a potential toxicant, scientists are also enthusiastic about the potential of chip technology and in vitro tests for asking specific questions—with data from human cells, rather than animal models of disease. "We can now go into more depth," says toxicologist Sandra Coecke of ECVAM. "With in vivo tests, you ended up with kind of a black box." Indeed, Coecke and others feel that these kinds of new methods—once validated—could not only replace animals tests, they could be an improvement.

# 7

# The Animal Rights Movement Threatens Medical Progress

## Frederick K. Goodwin

*Frederick K. Goodwin is a research professor of psychiatry at George Washington University and director of the university's Psychopharmacology Research Center. He also directs the university's Center on Neuroscience, Medical Progress & Society and is the former director of the National Institute of Mental Health.*

Actions taken by animal rights organizations to end biomedical research threaten to undermine medical progress. In fact, many important breakthroughs in the medical field, such as the use of lithium for the treatment of manic-depression, would not have occurred without the use of laboratory animals. Unfortunately, animal rights groups have helped create "a climate of moral confusion" that equates animal life with human life. Scientists have erred by giving in to their demands; in adopting "the three Rs" philosophy (reducing the number of animals used, refining research techniques, and replacing animals when possible), scientists appear to admit that animal testing is cruel and immoral, opening the door for greater demands from activists. As a result, the activists will not relent until animal experimentation has ceased.

The radical animal rights movement has become an increasingly powerful force that threatens the continued discovery and development of new treatments and prevention strategies for a variety of illnesses. The effort to end biomedical research with animals is based on a profound misunderstanding of how science really works and the gains we have achieved. Fundamentally, it represents a philosophical position reflecting a profound moral confusion that equates our use of animals with the enslavement of human being, and treats them as moral agents on a par with people.

An important truth about the history of medicine is that most major discoveries have come about by accident, often when a scientist has his

Frederick K. Goodwin, "How the Animal Rights Zealots Threaten Medical Progress," *Medical Economics*, vol. 77, March 6, 2000, pp. 217–21. Copyright © 2000 by Medical Economics Company. This article is adapted from a speech delivered at a recent Manhattan Institute forum. Reprinted with permission from the Manhattan Institute for Policy Research.

sights trained on an entirely different topic of research. Earlier in my career, I had the good fortune to be the first researcher to report on the antidepressant effects of lithium in a controlled study. But the story behind the initial discovery that lithium, an elemental substance on the periodic table, might have therapeutic benefits illustrates the serendipitous way medical advancement occurs.

An Australian psychiatrist named John Cade was interested in understanding what might be wrong in the brains of patients with manic-depressive illness. Because he was evaluating nitrogen metabolism, he wanted to see if giving them a substance called urea might help. Testing his hypothesis on guinea pigs, he used a salt form of urea which happened to contain lithium, in order to make a soluble solution, and gave it to the animals. What happened, of course, was that the guinea pigs became unexpectedly calm. Further experimentation revealed it wasn't the urea producing the effect, but the lithium, which came as a complete surprise to Cade and everyone else. He confirmed his findings by taking lithium himself and giving it to human patients, who experienced similar results. This single discovery has entirely revolutionized the treatment of manic-depressive illness, improving the lives of many people and saving billions of dollars along the way.

*I would challenge anyone to produce a working biomedical scientist who would dispute the importance of using laboratory animals.*

But there was no way anyone could have predicted what the outcome of this experiment would be in advance. And there was no way to list the health benefits that would come from using those guinea pigs before the experiment was done. It would have been like asking for an answer before the question was even clear. If you know the answer ahead of time, you're not really doing research.

## Scientists agree on lab animal use

Many years ago, well before the animal rights controversy arose, the National Institutes of Health sponsored a study to examine whether the government's funding of basic biomedical research was a good investment. As part of the study, the authors surveyed practicing cardiologists to determine what the group regarded as the 10 leading medical advances of their lifetimes—that is, the 10 developments most helpful to their patients. The authors then traced the scientific ancestry of each of these discoveries and found that, in every case, animal research was a critical component. And four out of the 10 actually originated from work in a different, seemingly unrelated field of research. So it's unreasonable to expect a scientist doing interesting work to know where his efforts may lead. We can evaluate whether an experiment is likely to give clear answers based on whether it is well designed, but we can't say what those answers are going to be.

And I would challenge anyone to produce a working biomedical scientist who would dispute the importance of using laboratory animals. Only about 22 percent of the work being done in biomedicine involves animals, more than 90 percent being rats and mice. But anyone working in the field will tell you that it's indispensable. You can't develop an understanding of a chemical or a gene and then try and determine its role in a complex human organism with billions of cells and dozens of organs without knowing how it works in the complex biological systems of animals. The animal model allows a scientist to understand what's happening at a level of detail that could not especially be achieved in humans.

There may be no more dramatic illustration of lab animals' importance than the work of the great kidney transplant pioneer, Dr. Thomas E. Starzl. When asked why he used dogs in his work, Starzl noted that in his first series of operations he transplanted kidneys in a number of subjects and the majority died. After figuring out what had allowed a few to survive, he revised his techniques, operated on a similar group of subjects, and the majority survived. In his third group only one or two died, and in his fourth group all survived. The point, he added, was that the first three groups were made up of dogs, while the fourth group consisted of human babies. If he had started on humans, he would have been responsible for 15 human deaths. Yet there are animal rights activists who believe that is the choice we should make.

To understand what the animal rights movement is all about, it's essential to distinguish between it and animal welfare organizations like the American Society for the Prevention of Cruelty to Animals (ASPCA) and local humane societies, which have a time-honored place in our culture. Typically these organizations try to reduce animal cruelty, take care of stray animals, teach good animal care, run neutering programs, and build animal shelters, thereby fulfilling our traditional moral responsibility as stewards of animals, who are not in a position to make decisions or care for themselves. Incidentally, my wife and I have two dogs and a cat, and have been supporters of our local humane society. Our pets come from their shelters.

## The philosophy of animal rights groups

The animal rights organizations, however, started with a very different philosophy, summed up by the grandfather of the movement, Professor Peter Singer. Singer has argued that all sentient creatures, all those capable of feeling, have the same fundamental rights. Anyone who assigns special rights to humans is guilty of "speciesism," a prejudice morally equivalent to racism and sexism. Ingrid Newkirk, a leader of the American animal rights movement, has written that "a rat is a pig, is a dog, is a boy." She has also compared 6 million Holocaust victims to 6 billion chickens killed. She also asserts that pet ownership is slavery. Chris Rose, who heads the organization In Defense of Animals, says if the death of one rat would cure all disease, it still wouldn't be right, because we are all equal.

Ironically, science itself may have indirectly played a role in creating an intellectual climate in which this kind of thinking is possible. Many people interpreted modern science to say that nothing exists except what we can measure empirically, and that if all truth depends on science, moral values are without foundation. This has contributed to a climate of moral, cultural, and intellectual relativism, a perspective that infuses postmodern thought so dominant in our most prestigious universities. But science itself absolutely depends on the concept that there is such a thing as truth and that there are systematic ways to distinguish truth from falsity.

## A climate of moral confusion

To be certain, other factors have contributed to the climate of moral confusion surrounding the use of animals in research, including the general erosion of trust in our institutions brought about by events like Vietnam and Watergate. We're also victims of our own health care successes. We have seen such a decline in infectious diseases that the baby boomers and subsequent generations don't even remember polio, and have no sense of how amazing it was when antibiotics were first developed.

And these days most people don't spend much time around animals other than house pets. Immediately after World War II, 44 percent of our population lived on farms; now it's around 4 percent. So what do kids know about animals, other than what they see in animated movies? And the animal rights movement tries very hard to attract young people. A recent television documentary on animal life generated a lot of protest just because it showed the reality of jungle life, namely that every animal is some other animal's dinner.

*Nothing destroys creativity like fear, and the [animal rights] movement has instilled a sense of fear that permeates the research community.*

Interestingly, animal rights activists made a decision early on to make scientific researchers their targets of protest rather than farmers, despite the fact that more than 99 percent of the animals used by humans are for food, while a fraction of 1 percent are used in research. Peter Singer said that decision was made because farmers are organized and politically powerful, and they also live out in rural areas, which makes them hard to get to. On the other hand, scientists are not politically organized, live in urban areas, and are not very good at defending themselves because they often have trouble explaining their work in layman's terms.

Scientists have indeed proved easy targets, partially because we made a very bad tactical error in the beginning. In an effort to meet the activists halfway, the research community came up with what were called "the three Rs." We were going to reduce the number of animals used, refine our techniques, and replace animals whenever possible. Subsequently we have had to relearn the old lesson that it is a losing game to compromise with a radical group that has an entirely different way of seeing the world.

For that reason, focusing on the three Rs without identifying the underlying philosophy of animal rights proved a public relations disaster. Our commitment to replacing animals suggested that we thought we might be doing something morally wrong. Our basic position should have been that human beings have a right to use animals for their purposes, but we also have responsibility to use them humanely. The more we emphasized the three Rs, the stronger the movement became, and the radical activists were able to raise more and more money.

## The weaknesses in activists' arguments

Animal rights organizations have advanced several core arguments, which I want to rebut:

• They assert animal research is essentially cruel. This argument misses the point that experimenters usually want to disturb the animal as little as possible in order to study its natural response to whatever is being tested. An estimated 5 percent of research employs procedures causing distress or pain. The reason is that the object of these studies is distress or pain. This kind of experimentation has allowed us to develop effective painkillers, for example. Moreover, animal research has become one of the most regulated forms of human endeavor.

• Activists say animal experiments are duplicative. The reality is that only about one out of four grant requests currently receives funding, a highly competitive situation that means that duplicative research is very unlikely to get funded. But research does have to be replicated before the results are accepted. And advancement usually comes with a series of small discoveries, all elaborating on or overlapping one another. When activists talk about duplication, they betray a fundamental misunderstanding of how science progresses.

• They say we should urge people to adopt measures such as an altered diet or increased exercise to prevent major illness, so that we wouldn't need so many new treatments. This misses the fact that much of what we have discovered about preventive measures is itself the result of animal research. You can't get most cancers to grow in a test tube; you need whole-animal studies.

• I'm amused most by the argument that we should use alternatives like computer simulations. I wonder where they think the data comes from that is then entered in computers? We have to use real physiological data to feed our machines. One activist I know keeps arguing we should use PET [Positron Emission Tomography] scans, which can provide an image of how a living human organ is functioning, as a way of avoiding the use of animals. He ignores the fact that it took Lou Sokoloff in my program at the National Institute of Mental Health eight years of animal research to develop the method upon which the PET scan is based.

• Despite the weakness of their arguments, the animal rights movement has already cost us a lot. Nothing destroys creativity like fear, and the movement has instilled a sense of fear that permeates the research community. The people who work with animals are now often segregated in high-security buildings like bunkers, separated from their colleagues. There has been an effective cut in biomedical research budgets resulting from the costs of increased security and compliance with new regulations. The public needs to understand what's at stake in this controversy before the costs mount higher.

# 8

# Using Animals as Organ Donors May Save Human Lives

## Daniel Q. Haney

*Daniel Q. Haney is a medical editor with the Associated Press news wire service.*

Despite some anticipated difficulties, researchers project that pigs may begin fulfilling the mounting human need for hearts and kidneys by 2006. In particular, genetically engineered miniature pigs, scaled down to the dimensions of a large person, show promise for supplying much-needed organs. Unlike primates, pigs are not endangered, and because their tissue is less like human tissue than the tissue of primates, their organs are less likely to successfully transmit a foreign virus during transplantation. Most importantly, many pig organs are similar in size and shape to human organs, making them more viable for transplantation.

As unlikely as it sounds, one solution to the shortage of organs for transplant could be the miniature pig, an animal that is already a lot like us and getting more so in the hands of genetic engineers.

The miniature pig weighs 300 pounds—one third the regular size—but otherwise is unmistakably all pig.

However, its relatively dainty dimensions have caught the eye of scientists, who note that it is about the size of a really large person. This means it is filled with nicely proportioned innards, especially a human-size heart and kidneys.

For this and other reasons, the pig is regarded as the most practical untapped source of needed body parts for sick and worn-out people. Perhaps 20 labs around the world are working to make pig parts fit for human transplants.

The goal: Clone and genetically modify pigs to "humanize" their organs. As that word implies, the animals are being changed in fundamental ways so they are less like pigs and more like people.

Researchers have already implanted some of these pig organs into baboons with modest success. Big scientific challenges loom, but [by 2006], if all goes well, the researchers hope to try these pig organs in people, offering redesigned pig hearts, kidneys and other organs to the desperately ill.

## Progress in xenotransplantation

The idea of transferring whole organs from animals to people has intrigued doctors for a century. The most famous patient, 12-day-old Baby Fae, received a baby baboon's heart in 1984. But like all such operations, that one ended in failure, and the infant died 20 days later when her body rejected the heart.

Those attempts were crude, compared with the current round of genetic manipulation and immunological tinkering by biotech firms, pharmaceutical companies and academic labs racing to make xenotransplantation a medical reality.

*The pig is regarded as the most practical untapped source of needed body parts for sick and worn-out people.*

If it works, the result will be limitless organs for human use. The idea hardly seems far-fetched to many transplant specialists, who watch thousands of patients die each year because of the shortage of human parts.

"I think it would be wonderful if we had a safe supply of organs that work as effectively as humans' (do)," says Dr. Patricia Adams of Wake Forest University, immediate past president of the United Network for Organ Sharing.

## The waiting list grows

According to the network, which manages the national transplant waiting list, about 77,000 Americans were in line for transplants [in 2000], while 23,000 actually received them. The waiting list is growing five times faster than the supply.

Those numbers understate the shortage. Because transplant rules are so strict, many who could benefit never make it onto the waiting list. For instance, hospitals generally will not consider heart transplant for anyone over age 65, no matter how healthy they otherwise are.

So without enough organs from cadavers, many researchers believe the best alternative is animals raised in germ-free barns near hospitals.

But pigs?

"Although it seems illogical, most people agree that the alternative species that makes the most sense is the pig," says Julia Greenstein, president of Immerge BioTherapeutics, a Boston company created [in 2001] to develop pigs for transplants.

Certainly, humans have nearer relatives that in some ways would be easier donors because their tissue is less foreign to the human body. For instance, organs taken from chimps could probably survive with nothing more than immune-suppressing drugs, but the animals are endangered, and many people would object to using humans' closest cousin for this purpose. While baboons are reasonably abundant, their organs are too small for adults.

Furthermore, taking organs from such closely related creatures could be risky. Apes and monkeys can carry viruses that are harmless to them but deadly to humans. The best example is the AIDS virus, which probably evolved in chimps.

So pigs' evolutionary distance from people is one argument in their favor. Of course, pigs have their own germs, and scientists take them seriously. But because pigs are so unrelated, the risk that their viruses would sicken people is thought to be slim.

Supply certainly is not a problem. Americans slaughtered 98 million pigs [in 2000]. And pigs' place on the food chain also probably means most people would not have ethical qualms about pigs for transplants.

"It is far more legitimate to have pig organs for human survival than pig meat for the supermarket. I think that's a slam dunk," says Harold Vanderpool, a bioethicist at the University of Texas Medical Branch who heads a xenotransplant advisory committee for the U.S Department of Health and Human Services.

But the biggest advantage of pigs is the striking similarity of many of their organs. For instance, their hearts are plumbed almost identically to people's. The pig kidney, lung, pancreas, and possibly even the liver appear similar enough to humans.

Some companies are concentrating on standard pigs, which reach about 1,000 pounds, on the theory that their organs will stop growing once they get large enough to keep a human alive. They note that a young rat's heart, when transplanted into a mouse, never grows to full size. But whether the same will be true for people with hearts from young, ordinary pigs is unknown.

The leading advocate of using the smaller organs of the miniature pig is Massachusetts General Hospital surgeon and immunologist David H. Sachs. "The miniature swine has the potential to donate an organ to any human being dying of organ failure," he says, "from a newborn baby to a sumo wrestler."

# 9

# Using Animals as Organ Donors Endangers Human Lives

## Stephanie Brown

*Stephanie Brown is a writer and animal rights activist living in Toronto, Canada.*

Transplanting organs from animals to humans, called xenotransplantation, would place the public at risk and should be banned. To date, every human recipient of an animal organ has died because their bodies have rejected the organs. Despite rigorous screening of the donor animal and monitoring of the recipient, there is still the danger that an unknown virus will be transmitted from an animal into the human population. Another consideration is the welfare of the animals; the lives of animal donors is abnormal and brief. While some scientists recognize the risks of xenotransplantation and have withdrawn support for it, animal organ suppliers and immunosuppression drug providers see the profit potential and are pressing for its commercial introduction.

A multinational drug firm and three Canadian universities, with government collaboration, are pushing a biotech experiment for Canada with potentially catastrophic health consequences.

The experiment is xenotransplantation, the transplanting of body parts from animals into humans. Preclinical animal experiments using transgenic pig organs in baboons are under way at the University of Western Ontario, in conjunction with the University of Toronto. The question is not if, but when, xenotransplants into humans will begin in Canada.

As attractive as xenotransplants may seem to people needing hearts, livers or kidneys, and to drug firms that stand to profit, the problems are many: Public health risks, massive financial costs, and profound ethical issues about our attitudes to animals.

Xenotransplantation experiments in Canada are funded by Novartis, the giant Swiss firm that dominates the billion-dollar global market for immunosuppression drugs required by organ recipients. Novartis stands to gain big if xenotransplants move forward.

Canada is not the only country experimenting with this technology. Animal experiments are under way in the U.S., New Zealand, Saudi Arabia, Germany, Sweden, Russia and India.

More human organs are needed than are available, so xenotransplants are seen as a solution. But to date, every human receiving an animal organ has died, with the longest surviving only a few months. High-profile cases include Baby Fae, an infant who received a baboon heart in 1984, dying 20 days later, and an AIDS patient who received a baboon liver in 1992. The problem is organ rejection: The immune system attacks the animal organ as foreign.

A persuasive argument against xenotransplants is the public health risk and the painful lesson of HIV [human immunodeficiency virus], which causes AIDS [acquired immune deficiency syndrome]. There are suspicions that HIV, a retrovirus that integrates its genetic material in the cells of the infected host, transferred to humans from monkeys. Pigs' genetic makeup, too, includes porcine endogenous retroviruses (PERVs), which cannot be screened out since they are part of the DNA. Once inside a host, they last forever. Sometimes called "stealth viruses," they can lurk for years or decades before causing symptoms.

The Canadian guidelines call for elaborate screening processes for known pathogens, but what of diseases not known? Two new PERVs have been discovered [since 1997], and there could be more. Test tube experiments have shown some viruses can be transmitted from a pig cell to a human cell. As xenotransplants increase, so will the danger of passing a virus to the wider population.

The guidelines recognize the uncertainty of screening animal organs for disease. They offer cautious words about disease transmission, such as "Life-long monitoring of (sentinel) animals will increase the probability of detection of subclinical, latent or late onset diseases."

[In 1999,] a new pig virus called Nipah, an encephalitis, killed 111 pig farmers in Malaysia when it spread from pigs to humans. Nearly a million pigs and dogs were slaughtered to control the outbreak.

---

*To date, every human receiving an animal organ has died, with the longest surviving only a few months.*

---

The University of Pittsburgh announced in September [1999] that tissues archived from the AIDS patient who received a baboon liver in 1992 contained a baboon virus, the first confirmation of a species-specific virus from a transplanted animal organ. It is now recognized baboons contain too many disease organisms to be suitable donors.

Who will accept liability if pig retroviruses enter the human population from xenotransplants? Would it be the government that allowed this experiment? Some scientists, perhaps recognizing their own possible complicity, have spoken out strongly against xenotransplants. Fritz Bach, a Harvard immunologist, said, "Xenotransplantation is a unique medical enterprise. It puts the public at risk for the benefit of the individual. If you

put the public at risk, then it has to be the public that says, 'I do not accept the risk, or I accept it'."

The push for xenotransplants has a strong commercial component.
The potential market for pig organs and associated pharmaceutical therapies is $11 billion per year, according to the *Wall Street Journal.*

There would be consequential costs with xenotransplants. Donated
human organs are free. Transgenic pig organs won't be, and it would be
a seller's market, ranging in the thousands of dollars per organ. Addi-
tionally, introduction of commercialized animal organs could harm human organ donation.

---

## *Who will accept liability if pig retroviruses enter the human population from xenotransplants?*

---

Significant costs, unique to xenotransplants, include: a registry to
track patients, "to manage the risk of xenotransplantation in Canada;" in-
definite archiving of multiple tissues from donor animal and organ recipient, including cryopreservation; lifetime monitoring of patients; and
oversight of patient caregivers and sex partners. All to "try to" prevent
transmission of new diseases.

In early November [1999], the United Kingdom announced standards
for xenotransplants into humans: Organ recipients may not have unpro-
tected sex and, therefore, may never have children. Patients must inform
authorities about each sex partner and those persons be identified to their
physician.

A safer alternative to animal organs is more human organs. Canada's
woeful record of organ and tissue donations—14.5 donors per million
population—is one of the lowest among industrialized nations. Inte-
grated strategies are needed to bolster human organ donation and pro-
curement.

Animals are half the xenotransplant equation, since their organs are
at stake. Human and nonhuman animals are alike in many ways, capable
of hunger and fear, pain and pleasure, comfort and anxiety. Despite this,
animals' rights are systematically violated.

True to form, the question of animal use is dismissed matter-of-factly
in the Canadian guidelines: "The adverse effects suffered by the pigs used
to supply organs for xenotransplantation would not outweigh the potential benefits to human beings." Easy enough for humans to decide.
Though guidelines call for "animal welfare," animal lives are given short
shrift, with the catch-all phrase "no unnecessary pain and distress," a
concept allowing virtually any procedure if humans dictate it's "necessary."

A common argument to justify pig organ use is the food issue. Yes,
millions of pigs are killed every year for food. It is argued, then, there is
no problem to kill pigs for their organs. Yet, if pigs weren't eaten, it would
prevent vast amounts of animal pain and be healthier for us. To defend
killing animals for one purpose based on another practice is not just.

Though pigs are curious, gregarious creatures, those bred to supply organs live in sterile conditions designed to minimize exposure to outside

pathogens. Their lives are unnatural and short. Piglets are caesarean-derived, fed by hand in incubators, not suckled by their mothers, at the University of Guelph [in Ontario, Canada]. Pigs "pharmed" for spare parts have been genetically engineered with the addition of a human gene researchers hope will overcome one aspect of organ rejection. The process of making a "new" transgenic animal is hit and miss. Besides eroding the integrity of the species, transgenesis is an uncertain process in which only one in 100 animals may carry the added gene, with the remaining 99 animals unwanted and wasted. Frankensteinian deformities can occur, with organs oversize or missing, or in wrong places.

Canadians have a right and responsibility to be heard on the introduction of untried biotechnologies. There is sufficient evidence now for a ban on xenotransplants. The risks are not just to organ recipients, but to us all.

# 10

# Experimentation on Nonhuman Primates Is Vital to Medicine

## Mick Hume

*Mick Hume is a columnist for the* Times *in London.*

Using nonhuman primates for research to benefit humans is right and proper and to suggest otherwise is ridiculous. Scientists could not have achieved advancements such as chemotherapy and antibiotics without experimenting on primates. Although there are some genetic similarities between nonhuman primates and humans, nonhuman primates cannot be considered equal to humans, despite the claims of animal experimentation opponents. The work of animal rights groups imperils the construction of new primate research facilities that will benefit humanity.

It is entirely moral, humane and proper to place electrodes in the brains of primates, as part of the search for a cure for such terrible diseases as Alzheimer's and Parkinson's. What is really sick is that so many seem to doubt it.

[In November 2002] a public inquiry opens into Cambridge University's plans to build a world-class, £24 million[1] research centre, where neuroscientists will experiment on the brains of primates. Two previous applications for approval for the institute have been turned down on spurious, non-scientific grounds, after complaints that it would infringe green belt planning regulations and endanger public order by provoking animal rights protests.

Whatever the outcome of the new hearing,[2] it is a disgrace that the Government feels the need to hold a public inquiry into whether we should privilege people or primates. The agenda on the animal research issue seems increasingly to be set by the lobby group Friends of the Planet of the Apes.

---

1. approximately 13.5 million in U.S. dollars in December 2003    2. In November 2003, Deputy Prime Minister John Prescott gave Cambridge University permission to proceed with its research center.

## The benefits of primate research

Animal research has been indispensable to the progress of medical science. We have all benefited from it, including every animal rights protester, unless they extend their ethical objections to never having taken a painkiller or antibiotics. And experiments on primates have played an important part in developing everything from chemotherapy to organ transplants. The use of primates is now widely considered unethical, on the grounds that they are "just like us" (speak for yourself). Yet it remains essential in important areas; for example, the current candidates for a vaccine against AIDS were all developed using primates.

*The agenda on the animal research issue seems increasingly to be set by the lobby group Friends of the Planet of the Apes.*

In an obscene inversion of the truth, opponents of the Cambridge institute now claim that its experiments could actually pose a threat to human health. Pushing panic buttons that invoke every scare from bioterror to BSE [Bovine Spongiform Encephalopathy or 'mad cow disease'], they warn that unidentified primate viruses could somehow escape from the research centre and spread exotic diseases among the local population (even the dread word "Ebola" has been whispered). No doubt we shall soon be told that Dolly the sheep was to blame for the anthrax panic.

## Manipulating the truth

In another brazen attempt to take a scalpel to the facts, anti-vivisectionists —who have long insisted that the genes we share with apes makes experimenting on them unethical—now argue that the Cambridge experiments are redundant because primates' brains are too dissimilar to ours after all. In truth it is the genetic similarity between humans and primates that makes experimenting on them expedient. And it is the qualitative difference between us and them in every other respect that makes such experiments ethical.

Opposition to animal research is an emotional spasm in search of a rational argument. Scratch the surface of much of this pseudo-science and you will find saccharine, Disney-style anthropomorphism. The website of the International Primate Protection League, a leading expert group campaigning against the Cambridge centre, prominently features little "Courtney Gibbon", a victim of maternal abuse now being hand-raised by IPPL [International Primate Protection League] "caregivers". IPPL members "who have made a special place for Courtney in their hearts" are kept informed of how she plays outdoors and "tries to sing along with the big gibbons in her tiny voice!" Why worry about all those old folks with Alzheimer's and Parkinson's when we've got Courtney the singing baby gibbon to coo over?

The people who conduct animal experiments are scientists, not sadists. But even if they had the morals of [convicted murderer] Myra Hindley, their work would still be worthwhile. If anything, the science community seems not hardhearted enough in defence of its work, often seeking to compromise with the critics rather than fighting to convince the public of a simple proposition: that animal welfare cannot be the business of animal research.

## Degrading human worth

The Cambridge initiative is in serious peril because the cause of animal rights is no longer the preserve of a few extremists. It has become mainstream in a British society where the traditional fondness for animals now appears to be married to a deep self-loathing of human achievements. If many now seem prepared to put primates on a par with people, it is not because we have discovered anything that raises the status of animals, but because we have lowered our estimation of our own moral worth.

The irony is that the argument for animal rights is itself an expression of thoughts and feelings that are uniquely human. Using our insight of self-consciousness, many now ask "what are we?", and don't like the answers that they are given. Some might seek to express their discomfort with human progress by denying our right to use animals, and especially primates, in the just cause of medical science. But even our capacity for self-loathing is a sign of humanity's superiority over all other species.

# 11

# Nonhuman Primates Should Not Be Used in Experiments

## Animal Aid

*Animal Aid, founded in 1977, is the United Kingdom's largest animal rights group.*

Nonhuman primates make poor subjects for experimentation, and their use may even cause harm to humans. For example, one drug for arthritis killed sixty-one people, despite having been tested successfully on primates. Infectious disease research on primates has encountered devastating failures as well, particularly in AIDS research. Similar problems have arisen in experiments on neurological disorders, including studies on stroke treatments. Since primate brains work differently than human ones, data from neurological studies on primates is invalid. Other forms of testing, such as magnetic resonance imaging (MRI) or *in vitro* (test tube) experiments, along with human-based studies, would produce more valid outcomes than do experiments on primates.

The public is strongly opposed to the use of primates in laboratories for a number of compelling reasons that cannot be dismissed as mere sentimentality. Many dispute the claim that research on primates is necessary for medical progress and believe that the reverse is true. As the following [viewpoint] will show, primates are a poor model for such research and their use has resulted in harm to patients, which is an inevitable consequence of reliance on other species to study human diseases.

Our close kinship with primates is undeniable and the more we learn about them, the more it becomes apparent that they share with us emotions, intelligence and complex social relationships. They are clearly capable of suffering psychologically as well as physically when separated from their family groups, confined in a cage, denied freedom to express their natural behaviour and subjected to painful and invasive procedures. All these fates await primates used in laboratory experiments. . . .

Animal Aid, "Monkeying Around with Human Health," www.animalaid.org.uk, June 29, 2003. Copyright © 2003 by Animal Aid. Reproduced by permission.

## Using primates in drug research

Primates have failed researchers with regard to their ability to predict dangerous side effects of medications. For example:

• Hormone replacement therapy—given to millions of women following research in monkeys—has recently been found to increase their risk of heart disease, stroke and breast cancer.

• Isoprenaline doses (for asthma) were worked out on animals, but proved too high for humans. Thousands of people died as a result. Even when the researchers knew what to look for they were unable to reproduce this effect in monkeys.

• Carbenoxalone (a gastric ulcer treatment) caused people to retain water to the point of heart failure. Scientists retrospectively tested it on monkeys, but could not reproduce this effect.

• Flosint (an arthritis drug) was tested on monkeys—they tolerated the medication well. In humans, however, it caused deaths.

• Amrinone (for heart failure) was tested on numerous nonhuman primates and released with confidence. People haemorrhaged, as the drug prevented normal blood clotting. This side effect occurred in a startling 20% of patients taking the medication on a long-term basis.

• Arthritis drug Opren is known to have killed 61 people. Over 3,500 cases of severe reactions have been documented. Opren was tested on monkeys without problems.

• Aspirin causes birth defects in monkeys but not in humans.

Despite these failures, marmosets, in particular, are increasingly popular as the 'second species'—in addition to rodents—required by regulators responsible for licensing new drugs. They are attractive to pharmaceutical companies because they are small and easy to breed in captivity. Their size makes them cheaper than dogs to dose with expensive test compounds and easy to house in small cages and inhalation chambers.

These benefits are itemised in a paper [written by D. Smith, et. al and] published by the Association of the British Pharmaceutical Industry in 2001. The paper notes, however, that marmosets are very excitable and can be difficult to handle. Their small size (and therefore blood volume) can be a problem when multiple blood samples are required. Skilled and experienced technicians are needed to dose marmosets intravenously, to take blood from their femoral (thigh) artery, or to dose them by 'gavage'—a long tube pushed down the throat to the stomach. Marmosets cannot be trained to tolerate these procedures and must be restrained and even sedated.

## Stolen from the wild

Not only do monkeys endure the trials of laboratory life, many are imported from such distant countries as Mauritius, Israel, Indonesia, the Philippines and China. 53% of procedures in 2001 involved animals imported from such sources outside the EU [European Union]. Investigations by the British Union for the Abolition of Vivisection . . . and by the RSPCA [Royal Society for the Prevention of Cruelty to Animals] . . . reveal appalling conditions at some breeding centres, which are often founded, re-stocked and augmented with animals trapped from the wild. Capture from the wild causes huge distress. The first-generation offspring are sold to UK [United Kingdom] laboratories, having been taken from their mothers as young as six months old. Their journeys to the UK are in tiny, cramped crates and can last as long as three days—some monkeys have died in transit.

---

*The use of primates—or indeed any animal species—for safety evaluation has never been scientifically validated.*

---

Concern about the use of macaques, in particular, is heightened by their conservation status. Long-tailed (also known as crab-eating or cynomolgus) macaques and rhesus macaques are the most commonly used species. They are 'old world' monkeys, native to Asia, where they live in large social troops that sometimes number 100 individuals. They are very communicative and maintain close relationships through mutual grooming. The long-tailed macaque is listed as near-threatened on the 2000 International Union for the Conservation of Nature 'red list'. The Japanese macaque is listed as endangered; yet up to 2,000 are captured and sold to Japanese laboratories every year. China is the main source of rhesus monkeys for Britain but housing conditions there are particularly horrifying. Breeding stock is taken from wild populations, which are in serious decline.

## Invalid testing

A fundamental point is that the use of primates—or indeed any animal species—for safety evaluation has never been scientifically validated. The effectiveness of the practice can be judged by the fact that, following the 'successful' completion of all the animal tests, more than 80% of new drugs fail when administered to healthy human volunteers during Phase 1 clinical trials.

There are more reliable methods to predict the safety and effectiveness of drugs for people. These include *in vitro* (test tube) studies using human cells and tissues, and sophisticated computer simulations designed to mimic human metabolism. A ten-year international study proved that human cell culture tests are more accurate and yield more useful information about toxic mechanisms than traditional animal tests. The British company Pharmagene uses human tissue exclusively, noting that 'a flood of new data on human genetics is making drug research in animals redundant. If you have information on human genes, what's the point in going back to animals?

Screening new drugs in silico (on computer) is now taking the place of many animal tests. German biotech company 4SC designs new drugs entirely in silico and can process in one day what would take other biotechs a month. 'The time is fast approaching when what we are doing will be the industry norm,' says chief executive, Ulrich Dauer. 'We have the accuracy, the speed and we don't waste time with drugs that are not going to work.'

The following example illustrates the ineffectiveness of assessing drug safety in animals and the impossibility of detecting subtle human responses: Eight out of ten drugs that were withdrawn from the US [United States] market between 1998 and 2001 had serious side effects in women that had not occurred in men. All of them had, of course, been tested extensively in animals before they were released onto the market.

If men cannot predict the effects of drugs for women, how on earth can we expect to obtain reliable data from monkeys?

## Infectious disease research

Investigating diseases that infect humans in any species other than humans is nonsensical, as pathogens and immune responses to them are highly species specific. For instance, chimpanzees are essentially immune to the human AIDS virus, Hepatitis B and C viruses, the malaria parasite and many other pathogens to which humans are susceptible.

The anthrax attacks in the US mail [in September 2001] were initially not taken seriously enough because experiments on monkeys showed the bacterium was not fatal until 8–10,000 spores are inhaled. When people died from much smaller doses it became apparent that this does not apply to humans.

---

*Far too frequently, animal models have been used to develop vaccines that are effective in laboratory animals but are ineffective, or actually harmful, in humans.*

---

The same failings apply to vaccine development:

Despite mounting evidence of vaccine research failures in animals, tens of thousands of primates and other animals have been killed in AIDS research over the past 20 years. This is despite the fact that infecting animals, even chimpanzees, with HIV [human immunodeficiency virus, which causes AIDS] does not produce an equivalent disease to human AIDS.

## Chimpanzee AIDS research abandoned

This reality has long been recognised by many in the research community and by AIDS activists, who have campaigned hard against futile vaccine research in monkeys.

After an extensive review of the American AIDS research programme, the US government concluded that chimpanzees are a deficient 'model' for use in AIDS research and redirected $10 million of funding. Even the director of the Yerkes Primate Centre admitted that 15 years of AIDS research in chimpanzees had produced little data relevant to humans.

Everything we know about HIV and AIDS has been learned from studying people with the disease—through epidemiology and *in vitro* research on human blood cells. Using primates to predict how humans will respond is not simply unproductive, it has resulted in medical catastrophe. In the early 1980s, the observation that HIV did not affect chimpanzees led scientists to assume that the virus would be harmless to humans too. They consequently advised health authorities to allow transfusions with contaminated blood samples, thereby giving rise to the French blood scandal that claimed thousands of innocent victims.

The first five-year trial of an HIV vaccine, 'Aidsvax', based on success in animals has [in 2003] been pronounced a failure. The 5,500 high-risk volunteers in the trial were not protected from HIV infection by the vaccine. Further trial results were due at the end of 2003. Many thousands of participants have been given false expectations which have been cruelly dashed.

Far too frequently, animal models have been used to develop vaccines that are effective in laboratory animals but are ineffective, or actually harmful, in humans. AIDS is a terrible illness, and research money and personnel need to be directed toward methodologies that are viable. Using an archaic methodology like animal models to combat a 21st century disease is more than foolish, it is immoral.

## The futility of primate brain research

Experimenting on monkeys with the hope of unlocking the secrets of the human brain is an exercise in futility. The most dramatic difference between humans and any other species, including the great apes, is found in the central nervous system. Our brain is four times larger than that of a chimpanzee, which is four times larger than that of a macaque. The human brain is enriched with specific cell types implicated in communication, language, comprehension and autonomic functions.

In addition to anatomical differences, the pattern of gene expression in our brain is dramatically different from that of the chimpanzee. Humans are distinguished from all mammals by their lack of a particular sugar molecule on the surface of cells, especially in the brain. It is likely that this profoundly affects brain development and function. Biochemical pathways in the human brain are unique.

Many of the attributes that we most celebrate—such as our ability to express ourselves in prose, poetry, song and dance—are uniquely human. We are clearly different, very different, from chimpanzees.

Yet at British universities, including Oxford, Cambridge, Manchester and London, monkeys are still used—at taxpayers' expense—as models of human brain function.

This is despite the fact that human brains can now be studied noninvasively using high-tec scanners. These enable the conscious brain to be observed while engaged in a variety of cognitive tasks (e.g. talking, singing, reading, writing) of which monkeys are not capable—and thus could clearly not provide any relevant insight.

## State-of-the-art research

Functional MRI [magnetic resonance imaging] scanners can monitor the brain activity of volunteers undertaking tests of memory and other skills, to reveal brain areas that are active during particular activities. Transcranial magnetic stimulation (TMS) temporarily disrupts brain function, allowing scientists to assess the impact of 'switching off' specific regions without permanently removing them. The Dr. Hadwen Trust for Humane Research is funding such studies into epilepsy research at Oxford University. There are many other state-of-the-art imaging techniques now available, including PET (positron emission topography), CAT (computer-aided tomography), MEG (magnetoencephalography), EROS (event-related optical signals) and VBM (voxel-based morphometric analysis). These remarkable techniques are able to differentiate such subtleties as musical ability or whether someone is lying or how hard they are concentrating. Insights that can be gleaned from monkeys seem absurdly crude by comparison.

---

*Valuable discoveries . . . from human-based research, render the study of artificial approximations of the disease in animals redundant.*

---

One study [in 2000] of macaque monkeys at Oxford University was aimed at determining the role of the cerebellum in cognition, by making a series of lesions in their cerebella. The monkeys' skulls were opened and 16 separate injections were made of an acid into the right hemisphere, followed a week later by further open-skull surgery and 16 injections into the left hemisphere. The animals were then tested, thousands of times, on cognitive tasks they had been trained to perform before their brain damage. Then they were killed and their brains extracted for analysis. The experiment served only to emphasise the difference between human and monkey brains, by contradicting similar studies that had already been conducted with brain-damaged human subjects. Self-evidently, the only way to investigate human brain function is to study the human brain. Results from the brains of any other species are simply misleading.

## Neurological disorder experiments

Increasingly, marmosets and macaques are being used to study neurological diseases such as stroke, Alzheimer's, Parkinson's and Huntington's. Monkeys do not suffer naturally from any of these conditions, so researchers destroy parts of their brain in order to generate superficially similar symptoms and then test potential treatments. But there are important differences between these naturally-occurring diseases in humans and the artificially-induced monkey versions—differences that render the monkey data invalid. For example:

  • Parkinson's disease becomes progressively worse in patients, while the chemically-induced marmoset version demonstrates gradual recovery
  • plaques and tangles in the brain are the hallmark of Alzheimer's disease in humans but not in monkeys
  • brain-lesioned marmosets used in the study of Huntington's disease do not replicate the pathology or symptoms of Huntington's disease
  • the cause of the brain damage is different and one would therefore expect the treatment to be different too
  • countless treatments for stroke have been developed in primates and other animals—yet all of them have failed or even harmed patients in clinical trials

## People provide the answers

Future advances in our understanding and treatment of neurodegenerative diseases will come from where they always have—human-based observation and ethical clinical research. Everything we know about these diseases has been learned from studying patients while they are alive and after they have died—as well as population research and studies using human tissues cultured from biopsies or from autopsy samples.

A new brain-imaging probe has allowed the visualisation of Alzheimer's plaques in the brains of living patients for the first time. This will enable earlier diagnosis and accurate monitoring of the effects of treatments. A number of genes implicated in both Alzheimer's and Parkinson's diseases have been discovered through population analysis. Biochemical pathways have been charted via the study of human brain tissue. It is now possible to keep slices of living brain tissue alive for weeks, allowing researchers to study the effect of chemicals on entire neural networks, not just individual cells. Tissues from different parts of the brain can be co-cultured on the same chip to examine the communication between them.

Population studies have demonstrated links between dementia and high cholesterol diets, as well as with smoking, inadequate vitamin B12 and folate intake, and low oestrogen levels. Valuable discoveries such as these, from human-based research, render the study of artificial approximations of the disease in animals redundant.

# 12

# Chemical Testing on Animals Saves Lives

## Gina Solomon

*Gina Solomon is a senior scientist with the Natural Resources Defense Council and an assistant clinical professor of medicine at the University of California, San Francisco.*

Testing chemicals on animals helps protect human health. For example, studies have indicated that frogs and rats suffer adverse effects from pesticides such as atrazine, diazinon, and Dursban. Animal test results led to the banning of these products by the Environmental Protection Agency. Without adequate testing, humans are at risk from unknown chemical agents in the environment. Whenever possible, animals should not be used for testing, but there are no alternatives available to test how chemicals might be related to birth defects or neurological and reproductive problems.

A few years ago, a pregnant woman came to see me in my clinic. She worked in a laboratory where she was exposed to a chemical solvent. She wanted to know whether the chemical might harm her fetus. A search of the data collected by the Environmental Protection Agency (EPA) quickly revealed that this chemical causes toxic effects in laboratory rats, resulting in fetal reabsorption. Although fetal reabsorption does not occur in humans, miscarriage does. Needless to say, I moved quickly to remove this woman from harm's way.

I do not relish the fact that chemicals are tested in animals, and for ethical reasons I did not participate in animal dissections in medical school because they were for practice rather than for protection of health and the environment. Tests in lab animals, however, can be critically important tools, along with non-animal tests and human epidemiologic studies, to protect people, pets, and wildlife from dangerous chemicals.

[In 2002], a researcher from the University of California at Berkeley published a study showing that tiny doses of atrazine, the most common pesticide in the United States and a major water contaminant, caused

Gina Solomon, "The Lesser Evil," *Earth Island Journal*, vol. 17, Autumn 2002, pp. 47–49. Copyright © 2002 by Earth Island Institute. Reproduced by permission of the author.

male laboratory frogs to become hermaphrodites, developing both testes and ovaries. NRDC (Natural Resources Defense Council) is petitioning EPA to ban this dangerous chemical, based significantly on the serious health risks to frogs.

People for the Ethical Treatment of Animals (PETA) has attacked NRDC because we support EPA's Endocrine Disruptor Screening and Testing Program. The program was created under the Food Quality Protection Act (FQPA), which Congress unanimously passed in 1996. The law aims to protect children from pesticides, including chemicals that disrupt the body's hormones and the endocrine glands that produce them.

---

*There is simply no non-animal alternative for tests searching for birth defects, neurological impairment, and reproductive problems.*

---

Since hormone and neurological systems in rats are very similar to those in humans, laboratory tests can yield invaluable information. For example, the notorious pesticide Dursban is off the shelves of hardware stores because it was found to impair brain development in laboratory rats. Since Dursban was used for flea control on dogs and cats, the ban protected both pets and children. [In 2001] the EPA banned diazinon for household use for the same reasons. The endocrine disrupting pesticide vinclozolin is no longer on the fruit we eat because it was found to cause deformed penises in laboratory rats. PETA misleads with their assertion that EPA has not banned chemicals using the Toxic Substances Control Act. In fact, many chemicals have been banned or controlled under a wide array of laws, by numerous federal, local, and state agencies because of toxic effects that appeared in lab animals.

What are all the other harmful chemicals that we are routinely exposed to? The fact is we don't know. However, we do know that literally thousands of chemicals are being released into our air and water or sold in consumer products despite utterly inadequate assessments of their safety. Among nearly 3,000 chemicals produced at over a million pounds per year in the United States, less than a quarter have been tested for chronic (long-term, or cumulative) health effects. This is disgraceful. It means that all of us—our children, pets, and wildlife—are guinea pigs in a huge uncontrolled chemical experiment.

## No adequate alternatives

NRDC would prefer not to subject any animals to testing. But the alternatives—continued ignorance or human testing—are unacceptable. There is simply no non-animal alternative for tests searching for birth defects, neurological impairment, and reproductive problems. Even where non-animal tests exist, it is often impossible to extrapolate the results to humans.

Animal testing should be minimized or eliminated when scientifically appropriate, and the welfare of test animals must be a central concern of any testing program. NRDC recently negotiated a legal settlement with EPA in which the agency agreed to reduce the number of animals used in the endocrine disruptor program, refine procedures to make the tests less painful or stressful, and replace animals with non-animal systems when scientifically appropriate.

If PETA succeeds in paralyzing EPA toxicology programs, the winners will be the major chemical and pesticide companies. The industry would love to manufacture and profit from chemicals without worrying that the public will find out its products may cause serious health effects. The chemical manufacturers would love not to worry about EPA using scientific information to tighten regulations or even to ban their products.

We need all the information and all the tools that we can muster in order to prevent harm from the thousands of chemicals that are used in our workplaces, schools, and consumer products, and that are being released into our air and water and spread on our food. While we would prefer not to sacrifice a single laboratory rat, we believe that the sacrifice is warranted to protect our children and future generations.

# 13

# Chemical Testing on Animals Is Unreliable

## Alix Fano

*Alix Fano is director of the Campaign for Responsible Transplantation and the author of* Lethal Laws: Animal Testing, Human Health and Environmental Policy.

While mice are regularly used in chemical testing, their physiology—which is very different from humans—makes them inadequate and unreliable subjects. For that reason, results from chemical tests on mice can be misleading. Furthermore, the experiments are extremely painful for the animals and are undermined by the use of improper test methods. For example, the chemical doses fed to mice are higher than any dose a human would likely be exposed to. Moreover, testing conditions are stressful, which tends to skew the results because stressed animals react differently to substances than do relaxed animals. The fact that pollution and unsafe chemicals still exist in the environment indicates that these experiments have not helped to safeguard human health.

There is a creature that lives two to three years, is unable to vomit, has no gall bladder, will give birth to 100 young each year, can synthesize Vitamin C in its body, and could be up to three billion times more cancer-prone than a human.

That creature is a mouse, and it is used for scientific research into finding a cure for cancer in humans.

Here is some more rodent information. Just as people react differently to chemicals depending on various factors, animal test results vary widely according to the species, sex, age, diet, stress level, and strain of the animal. For example, N2-fluorenylacetamide has caused bladder cancer in male and female Slonaker rats, liver cancer in male, and breast cancer in female Wistar rats, and intestinal cancer in male and female Piebald rats. Benzidine has caused bladder cancer in humans and dogs, liver and mammary tumours in rats.

The apparently obvious conclusion from this is that laboratory animals do not have anything like the same biology as humans. Yet although animal toxicity tests have never been scientifically validated to determine whether they can effectively predict toxicity for humans, a mind-boggling array of animal-based data now fills toxicology manuals, textbooks, and computer databases. US regulators have used these data to establish environmental health standards through the Food, Drug and Cosmetic Act of 1938, the (now defunct) Delaney Act of 1958, the Clean Air Act of 1970, the Safe Drinking Water Act of 1974 and the Food Quality Protection Act of 1996.

## The studies on arsenic

Some of the conclusions drawn from animal experimentation can make quite horrific reading. Take, for example, the studies on arsenic and its potential for causing cancer. While numerous epidemiological studies have proven that arsenic causes cancer in humans, toxicologists now acknowledge that arsenic rarely, if ever, causes cancer in animals. Rats, for example, are remarkably resistant to the chemical and develop none of the illnesses—liver, bladder, kidney, and skin cancer—observed in humans. Only when researchers have gone to great lengths—implanting high doses of arsenic compounds in rats' stomachs, under the skin of newborn mice, and into the tracheas of hamsters—have stomach and lung cancers eventually been produced.

Animal tests with arsenic began in 1911 and are still ongoing today. Why? Have they prevented humans from being exposed to arsenic? In fact, no.

In February 2000, the Natural Resources Defence Council (NRDC), an environmental advocacy group, released a report which revealed that tens of millions of Americans have been drinking water containing unsafe levels of arsenic for decades. But arsenic is not the only concern.

## Chemical testing methods

There are currently 85,000+ chemicals on the market—dyes, insecticides, fungicides, herbicides, rodenticides, soaps and detergents, synthetic fibres and rubbers, glues and solvents, paper and textile chemicals, plastics and resins, food additives and preservatives, refrigerants, explosives, chemical warfare agents, cleaning and polishing materials, and cosmetics—and 1,500–2,000 new chemicals are added to that toxic flow each year.

Government agencies, such as the US Environmental Protection Agency (EPA), have set up massive animal testing programs, using mostly rats, mice, guinea pigs, rabbits, dogs, cats, hens, and fish, to allegedly test the safety of these chemicals.

In the tests, animals are forced to eat and drink chemicals using such crude methods as gavage—whereby a tube is surgically inserted into the stomach; they are forced to inhale toxin vapours, have chemicals injected into their bodies, painted on their skin, and dropped in their eyes. In reproductive studies, pregnant animals are fed chemicals and induced to abort their young; rats and rabbits will have their entire uteruses removed before expected delivery dates so their foetuses can be weighed, and dissected. The EPA still conducts the Lethal Dose 50 (LD50) test—recording how much chemical kills 50 per cent of the animals in a test group—even though a majority of scientists agree that the test is a lousy predictor of human risk. In these tests, animals suffer convulsions, severe abdominal pain, seizures, tremors, and diarrhoea. They bleed from their genitals, eyes and mouth, vomit uncontrollably, self-mutilate, become paralysed, lose kidney function, and fall into comas. Up to 2,000 animals may be killed in these ways to test just one chemical.

## The problems with testing methods

Animals are typically tested using methods and doses that are at odds with real-life conditions. In one experiment involving the sweetener cyclamate, animals were given the human equivalent of 552 bottles of soft drinks a day. In two experiments with trichloroethylene, used as a decaffeinating agent in coffee, rats were given the human equivalent of 50 million cups of coffee a day. Herman Kraybill of the National Cancer Institute has stated that such high dosing can falsify an experiment in two ways: it can either poison the cells and tissues so severely as to prevent a carcinogenic response that might otherwise have occurred, or it can so overload and change metabolic processes as to cause a carcinogenic response that might not have occurred. The reasoning behind dosing animals with quantities of chemicals that are irrelevant to natural human (or animal) conditions is that these methods will more reliably produce acute toxic effects, including tumours, in statistically significant numbers. But the majority of humans do not die from acute poisoning. Rare toxicities are what kill a lot of people, and these could be detected in tightly monitored human studies.

Data from animal tests are also influenced by the method chosen to expose the animal to a chemical. In one study with methylene chloride, chronic inhalation studies produced increased lung and liver tumours in rodents, whereas a drinking study failed to produce any tumours. Ironically, putting doses of test chemical in food or water is one of the more common methods used by toxicologists to expose animals to chemicals. But rats readily associate food with illness and will avoid a food if they have been ill after eating it. How much an animal eats or drinks—as well as the animal's age, genetics, and metabolism—can influence the outcome of an experiment.

*Although animal toxicity tests have never been scientifically validated . . . a mind-boggling array of animal-based data now fills toxicology manuals, textbooks, and computer databases.*

Some scientists claim that animal studies have shown how compounds like hormones can increase the risk of cancer in animals. But they fail to mention that circulating levels of oestrogen and progesterone differ as much as three-fold between rodents and humans. Veterinarians have seen elevated hormone levels in rabbits for up to 24 hours after the animals were moved from one room to another.

Stressful laboratory conditions and controversial dosing practices call into question the value of animal-to-human extrapolations and the vast databases of animal toxicity data. It has also been noted that the artificial laboratory environment, with its cold metal cages, sterilised food, water, and bedding, fluorescent lighting, temperature controls, and the pain of experimentation, is so stressful for the animals as to be causing them to develop cancer and other effects which would not be observed outside the laboratory.

## New testing programs

Beginning in early October 1998, the EPA announced three controversial animal-based toxicity testing programmes. The High Production Volume (HPV) Chemical Challenge programme calls for toxicity testing of 2,800 chemicals imported or manufactured in amounts of one million pounds per year. Devised in closed-door meetings between the EPA, the Chemical Manufacturers Association, and the Environmental Defence Fund without public notice or Congressional oversight, the programme is slated to cost $700 million to implement, $11 million to administer, and would poison some 1.3 million animals.

In the second programme, the Endocrine Disruptor Screening Programme (EDSP), sixty thousand chemicals are to be tested on tens of millions of animals to determine whether and how chemicals disrupt the human hormonal system, despite crucial differences in humans' and animals' endocrine systems. Circulating levels of oestrogen and progesterone differ as much as three-fold between rodents and humans, as mentioned before; and reproductive geneticist Jimmy Spearow found that the CD-1 mouse strain, favoured by toxicologists because it produces large litters, is 16 times more immune to the effects of endocrine disrupting chemicals than other mouse strains. Toxicologist John Giesy, a member of the National Academy of Sciences, has said it is 'unbelievably stupid and a waste of resources' to legislate endocrine testing given the high level of uncertainty surrounding the endocrine disruption theory.

---

*Animals are typically tested using methods and doses that are at odds with real-life conditions.*

---

At least 200,000 animals are slated to be killed in the third programme, the EPA's Child Health Testing Programme (CHTP) which requires 10 separate animal tests for each chemical (including the barbaric LD50 test) to allegedly 'assess the special impacts of industrial chemicals on children'. The EPA has refused to disclose the list of chemicals it plans to test, perhaps fearing the same sort of criticism it received for its HPV and endocrine disruptor programmes.

## Setting 'safe' doses

Various groups, including the People for the Ethical Treatment of Animals and the Physicians' Committee for Responsible Medicine (PCRM), have proposed an alternative to the EPA's animal testing plan. It would require the agency to take concrete action to eliminate or reduce chemicals—like lead, mercury, and pesticides—already known to be highly toxic to children. The agency has refused to consider the proposal, preferring instead to focus on setting 'safe' doses of chemicals for children in air, water, food, and breast milk.

But a pamphlet published by the National Cancer Institute states, 'there is no adequate evidence that there is a safe level of exposure for any carcinogen[ . . . ] In addition, a low exposure that might be safe for one person might cause cancer in another'. Some people are chemically intolerant as evidenced by a condition called Multiple Chemical Sensitivity, which affects about 30 per cent of Americans. So how does the EPA establish allegedly 'safe' doses of chemicals for children, or for anyone?

---

*Animal testing is . . . commercialised exploitation in its most brutal form.*

---

One EPA document explains: 'To predict the risk [of cancer] for humans, the oral doses used in animal studies are corrected for differences in animal and human size and surface area which has been accounted for by the cube root of the ratio of the animal to human weight'. Essentially, animal data are churned through complex mathematical formulae, adjusted by some arbitrary numerical factors, and voilá—out come the 'Maximum Contaminant Levels', 'Acceptable Daily Intakes' and 'Permissible Exposure Levels'—numbers representing the amount of chemicals in air, water, and food that a human can ingest over a lifetime with allegedly little risk of becoming ill.

In reality, weak environmental laws and lax enforcement ensure that animal-derived safety standards are ignored. We are surrounded by pollution which animal tests have clearly failed to prevent. Chemicals are left on the market, or used illegally even after they have caused cancer and other effects in humans and animals, rendering the existing safety standards for chemicals useless.

More importantly, all safety standards are set for individual chemicals and ignore that we are all exposed to thousands of chemicals in combination in our air, water and food.

It seems unlikely that we will ever determine the cumulative effects of chemical pollution. Tests on non-humans will merely add layers of complexity and confusion to what is already an uncertain process.

## The profit in animal testing

As in so many other branches of public policy, money is power and 'governments have a habit of backing the ideas of whoever pays the most tax'. Clearly, companies have relied on animal testing programmes to make chemicals acceptable to regulators, attractive to consumers—and to protect themselves from costly litigation. Cost benefit decisions have ensured that most chemicals remain on the market, regardless of whether they have caused cancer and other effects in humans and animals. Animal testing has become the principal component of a regulatory system that, in making concessions to industry, has lost sight of its mandate to protect human health. And humans have become the ultimate 'guinea pigs' in an increasingly polluted world.

Unfortunately, a number of mainstream environmental groups, such as the Environmental Defense Fund and Greenpeace, still believe that animal tests have effectively protected public health and the environment and led to chemical bans.

But the array of chemicals that humans have become exposed to, since animal testing programmes were institutionalized in the 1920s, has grown exponentially; and only a small handful of chemicals have ever been banned. Some of these, like DDT, and DES, continue to be used illegally; and mixtures of carcinogens, like PCBs and dioxins, persist in our environment with unknown consequences.

On a philosophical level, animal testing is part of the same life-destroying paradigm that environmentalists claim to oppose. It is commercialised exploitation in its most brutal form. Exposing tens of millions of animals, dozens of animal species, to unimaginable pain and suffering under the pretext of protecting public health, while simultaneously allowing the continued production and release of thousands of poisons into the environment, is unethical and unsound public policy.

# Organizations to Contact

The editors have compiled the following list of organizations concerned with the issues debated in this book. The descriptions are derived from materials provided by the organizations. All have publications or information available for interested readers. The list was compiled on the date of publication of the present volume; names, addresses, phone and fax numbers, and e-mail addresses may change. Be aware that many organizations take several weeks or longer to respond to inquiries, so allow as much time as possible.

**American Association for Laboratory Animal Science (AALAS)**
9190 Crestwyn Hills Dr., Memphis, TN 38125-8538
(901) 754-8620 • fax: (901) 753-0046
e-mail: info@aalas.org • website: www.aalas.org

AALAS is dedicated to the humane care and treatment of laboratory animals and the research that leads to scientific gains that benefit humans as well as animals. The association provides a forum for the exchange of information and expertise regarding the care and use of laboratory animals for clinical veterinarians, technicians, researchers, and educators. AALAS publishes two bimonthly journals: *Comparative Medicine*, a journal of comparative and experimental medicine, and *Contemporary Topics*, focusing on AALAS business. It also publishes *Tech Talk*, which presents information about laboratory animal science.

**Animal Legal Defense Fund (ALDF)**
127 Fourth St., Petaluma, CA 94952-3005
(707) 769-7771 • fax: (707) 769-0785
e-mail: info@aldf.org • website: www.aldf.org

Founded in 1979, the ALDF works within the U.S. legal system to end the suffering of abused animals. Its website provides updated news on animal use as well as ways to become involved in the cause for animal rights. Members may receive a newsletter.

**Animal Welfare Information Center (AWIC)**
10301 Baltimore Ave., 4th Floor, Beltsville, MD 20705-2351
(301) 504-6212 • fax: (301) 504-7125
e-mail: awic@nal.usda.gov • website: www.nal.usda.gov/awic

AWIC, an arm of the U.S. Department of Agriculture, provides information for improving animal care and use in research, teaching, and testing through the publication of newsletters, information guides, bibliographies, and fact sheets. It also provides updated news on animal research.

**Foundation for Biomedical Research (FBR)**
818 Connecticut Ave. NW, Suite 200, Washington, DC 20006
(202) 457-0654 • fax: (202) 457-0659
e-mail: info@fbresearch.org • website: www.fbresearch.org

FBR seeks to promote public understanding and support for the humane and responsible use of animals in medical and scientific research. Through its education programs, FBR informs the news media, teachers, students, and the general public about the need for animals in medical and scientific research and advancement. Its website provides links to other organizations as well as full texts of the Animal Welfare Act and the various policy manuals on handling laboratory animals published by the U.S. government.

**Great Ape Project (GAP)**
917 SW Oak St., Suite 412, Portland, OR 97205
(503) 222-5755 • fax: (503) 238-5884
e-mail: gap@greatapeproject.org • website: www.greatapeproject.org

The goal of GAP is to achieve for the great apes the same basic moral and legal protection that human beings have. Its website provides updated news events, ways to become involved, links to other sites, and a free newsletter.

**Humane Society of the United States (HSUS)**
2100 L St. NW, Washington, DC 20037
(202) 452-1100
website: www.hsus.org

The Humane Society is dedicated to creating a world where animals are treated with compassion and are respected for their intrinsic value. The website offers news on pets, wildlife, farm and marine animals, and animals used in research. Workshops and educational programs are available. HSUS also publishes *All Animals*, a quarterly magazine with photos and stories about animals.

**National Anti-Vivisection Society (NAVS)**
53 West Jackson Blvd., Suite 1552, Chicago, IL 60604
(800) 888-NAVS (6287) • (312) 427-6065
e-mail: feedback@navs.org • website: navs.org

The NAVS seeks to abolish the use of animals in research, education, and product testing. Its educational programs are directed at increasing public awareness about vivisection, identifying humane solutions to human problems, and developing alternatives to the use of animals in experimentation. The website provides a brief history of vivisection. Members receive updated news regarding the field of animal testing.

**National Association for Biomedical Research (NABR)**
818 Connecticut Ave. NW, Suite 200, Washington, DC 20006
(202) 857-0540 • fax: (202) 659-1902
e-mail: info@nabr.org • website: www.nabr.org

NABR is a national nonprofit association that advocates public policy recognizing the vital role of humane animal use in biomedical research, higher education, and product safety testing. It publishes news updates, presents conferences, and represents members at all levels of government.

**Office of Laboratory Animal Welfare (OLAW)**
**National Institutes of Health**
RKL1, Suite 360, MSC 7982, 6705 Rockledge Dr., Bethesda, MD 20892-7982
(301) 496-7163 • fax: (301) 402-2803 • fax: (301) 402-7065
e-mail: olaw@od.nih.gov • website: http://grants1.nih.gov/grants/olaw.htm

OLAW develops, monitors, and exercises compliance regarding the Public Health Service's policy on the humane care and use of laboratory animals involved in research conducted or supported by any component of the Public Health Service. OLAW provides training materials for working with nonhuman primates, dogs, and rodents in the laboratory as well as a manual for all laboratory animal care.

**People for the Ethical Treatment of Animals (PETA)**
501 Front St., Norfolk, VA 23510
(757) 622-PETA (7382) • fax: (757) 622-0457
e-mail: info@peta.org • website: www.peta.org

PETA is an international nonprofit organization based in Norfolk, Virginia, that operates under the principle that animals should not be eaten, worn, experimented on, or used for entertainment. PETA educates policy makers and the public about animal abuse through lobbyists, protesters, advertising materials, and its national and international websites.

# Bibliography

## Books

Ruth Ellen Bulger et al. *The Ethical Dimensions of the Biological and Health Sciences.* New York: Cambridge University Press, 2002.

Carl Cohen and Tom Regan *The Animal Rights Debate.* Lanham, MD: Rowman & Littlefield, 2001.

Pietro Croce *Vivisection or Science?: An Investigation into Testing Drugs and Safeguarding Health.* New York: Zed Books, 1999.

Kevin Dolan *Ethics, Animals, and Science.* Malden, MA: Blackwell Science, 1999.

Abul Fadl Mohsin Ebrahim *Organ Transplantation, Euthanasia, Cloning, and Animal Experimentation: An Islamic View.* Leicester, England: Islamic Foundation, 2001.

John P. Gluck et al. *Applied Ethics in Animal Research: Philosophy, Regulation, and Laboratory Applications.* West Lafayette, IN: Purdue University Press, 2002.

Lesley Grayson *Animals in Research: For and Against.* London: British Library, 2000.

C. Ray Greek and Jean Swingle Greek *Sacred Cows and Golden Geese: The Human Cost of Experiments on Animals.* New York: Continuum, 2000.

Lynette A. Hart *Responsible Conduct with Animals in Research.* New York: Oxford University Press, 1998.

Jann Hau and Gerald L. Van Hoosier Jr. *Handbook of Laboratory Animal Science.* Boca Raton, FL: CRC Press, 2003.

Vaughan Monamy *Animal Experimentation: A Guide to the Issues.* Cambridge, England: Cambridge University Press, 2000.

F. Barbara Orlans et al. *The Human Use of Animals: Case Studies in Ethical Choice.* New York: Oxford University Press, 1998.

Ellen Frankel Paul and Jeffrey Paul *Why Animal Experimentation Matters: The Use of Animals in Medical Research.* New Brunswick, NJ: Social Philosophy and Policy Foundation, 2001.

Bernard E. Rollin *The Unheeded Cry: Animal Consciousness, Animal Pain, and Science.* Ames: Iowa State University Press, 1998.

Deborah Rudacille *The Scalpel and the Butterfly: The War Between Animal Research and Animal Protection.* New York: Farrar, Straus, & Giroux, 2000.

Deborah J. Salem and Andrew N. Rowan *The State of the Animals 2001.* Washington, DC: Humane Society Press, 2001.

74

Niall Shanks — *Animals and Science: A Guide to the Debates.* Santa Barbara, CA: ABC-Clio, 2002.

Colin Spedding — *Animal Welfare.* Sterling, VA: Earthscan Publications, 2000.

Steven M. Wise — *Drawing the Line: Science and the Case for Animal Rights.* Cambridge, MA: Perseus Books, 2002.

## Periodicals

Jim Adams — "A Look Inside an Animal Research Lab," *Star Tribune* (Minneapolis, MN), October 13, 1999.

Anil Ananthaswamy — "Quality of Life: Home Comforts for Lab Animals Create Problems for Researchers," *New Scientist,* March 9, 2002.

Andrew Bernstein — "Animal-Rights Movement Seeks Not to Prevent Needless Cruelty to Animals, but to Inflict It upon Human Beings," Knight Ridder/Tribune News Service, April 19, 2000.

Rick Bogle — "Primate Annihilation," *Animals' Agenda,* May/June 1999.

J. Bottum — "The Pig-Man Cometh," *Human Life Review,* Winter 2001.

Nell Boyce — "Ouch! That Hurts," *New Scientist,* September 9, 2000.

Jerome Burne — "Comment & Analysis: Animal Testing Is a Disaster," *The Guardian,* May 24, 2001.

Diane Clay — "Decades of Medical Testing Now Paying Off for Animals," *Daily Oklahoman,* March 18, 2003.

Andy Coghlan — "Is It Less Cruel If the Animals 'Want' to Do It?" *New Scientist,* July 6, 2002.

Rodger D. Curren and Erin H. Hill — "From Inhumane to In Vitro: The Changing Face of Science," *Animals' Agenda,* November/December 2000.

Alix Fano — "Xenotransplantation Is Not the Solution to the Organ Shortage," Knight Ridder/Tribune News Service, February 24, 1999.

Jon Ferry — "New Calls Made for Changes in Research on Animals," *Lancet,* December 4, 1999.

Douglas Foster — "Open the Labs and Set Them Free?" *Los Angeles Times Magazine,* June 2, 2002.

Robert Garner — "Animal Rights and Wrongs," *Chemistry and Industry,* January 4, 1999.

Eduardo Goncalves — "Lambs to the Slaughter," *Ecologist,* March 2002.

Henry E. Heffner — "The Symbiotic Nature of Animal Research," *Perspectives in Biology and Medicine,* Autumn 1999.

Jonathan Hughes — "Xenografting: Ethical Issues," *Journal of Medical Ethics,* February 1998.

John Leo — "Another Monkey Trial," *U.S. News & World Report*, September 20, 1999.

Keri Lipperini — "Animal Research," *Paraplegia News*, October 2002.

Paul McCartney — "'Some Animal Tests Are Needed . . . but It Was So Difficult for My Linda'; Sir Paul Tells DES of Dilemma Over Wife's Cancer Drugs," *Mirror* (London), October 23, 1998.

Sarah Rose A. Miller — "Animal Research," *Humanist*, September 2001.

Adrian R. Morrison — "Personal Reflections on the 'Animal-Rights' Phenomenon," *Perspectives in Biology and Medicine*, Winter 2001.

David S. Oderberg — "The Illusion of Animal Rights," *Human Life Review*, Spring/Summer 2000.

Lawrence Osborne — "Fuzzy Little Test Tubes," *New York Times Magazine*, July 30, 2000.

Ellen Frankel Paul — "Why Animal Experimentation Matters," *Society*, September/October 2002.

Scott Plous and Harold A. Herzog — "Poll Shows Researchers Favor Lab Animal Protection," *Science*, October 27, 2000.

Ian Roberts et al. — "Does Animal Experimentation Inform Human Healthcare?" *British Medical Journal*, February 23, 2002.

Vicky Robinson — "Do We Have the Right to Play God?" *Express on Sunday* (London), January 6, 2002.

Jessica Sandler — "PETA Says No to Testing," *Earth Island Journal*, Autumn 2002.

Roger Scruton and Andrew Tyler — "Do Animals Have Rights?" *Ecologist*, March 2001.

Brandon Spun — "Is Animal Research Really Necessary?" *Insight on the News*, June 24, 2002.

Jerrold Tannenbaum — "The Paradigm Shift Toward Animal Happiness," *Society*, September/October 2002.

Alex Tizon — "Animal Rights Activists Want Great Apes Recognized as People Too," *Seattle Times*, March 29, 2000.

Frankie L. Trull — "Limiting Animal Research Would Be Cruelty to Humans," *Insight on the News*, July 29, 2002.

Pamela S. Turner and Kathiann M. Kowalski — "Are Ape Rights the Next Frontier?" *Odyssey*, October 2001.

John Vane — "Truth, Unlike a Meal, Can Be Hard to Swallow," *New York Times Higher Education Supplement*, December 21, 2001.

Steven Wise — "Beastly Behavior? A Law Professor Says It's Time to Extend Basic Rights to the Animal Kingdom," *Washington Post*, June 5, 2002.

# Index

acyclovir, 23–24
advocacy groups, 62, 63, 66, 70
AIDS (Acquired Immune Deficiency Syndrome),
  animal testing on
  abandonment of chimpanzees and, 7, 58–59
  breakthroughs in, 22, 24
  dependency on primates for, 6, 7, 53
  high failure level of, 25, 26, 55, 58
  organ transplantation and, 47, 49
AIDS vaccine, 53, 58, 59
air contamination, 66, 69
Alzheimer's disease, 52, 53, 60, 61
Americans for Medical Advancement, 25
American Society for the Prevention of Cruelty
  to Animals (ASPCA), 42
Animal Aid, 55
animal cruelty
  charges of, against McDonald's, 17
  extent of suffering and, 6, 13, 29–30
  fashion-industry and, 11, 16
  laboratory cases of, 28–29, 31–32
  organizations for prevention of, 42
  researcher responses to, 32–33, 42
  in ridiculous studies, 30
  in "standard four" tests, 31
animal experimentation, 6–9, 23, 33
  fundamental flaws in, 7
  is becoming more humane, 34–39
  is cruel, 19, 28–33
    con, 34–39
  is essential for medical research, 22–24
    con, 25–27
  legal protection from, 17, 18
  moral debate over, 14–15
  replacement strategies and, 17, 23, 35, 36,
    43–44
  see also ethics
*Animal Liberation* (Singer), 21
animal rights
  advocates, 8–9, 17, 55
  Christian perspectives and, 18, 19
  as entitled, 10–17
    con, 18–21
  history of philosophies, 19–20, 42–43
  human rights vs., 10–11, 18
  legal perspectives and, 8–9, 17, 18
  lobby groups and, 52, 53
  moral principles and, 7–8
  public attitude toward, 17, 43, 52
  reciprocity of, 11
  university courses and, 18
animal rights movement
  degradation of humans and, 19, 54
  legal basis for, 18–19

moral confusion surrounding, 43–44
  philosophy of, 19, 42–43
  as radical force, 6, 9, 40, 44
  scientists views vs., 40–42
  threatens medical progress, 40–44
  in United Kingdom, 52, 53, 55
  weak arguments of, 44, 54
animals
  humans as similar to, 8, 50, 55
    con, 54, 65–66
  humans as superior to, 8–9, 19, 54
  as organ donors
    endanger human lives, 48–51
    may save human lives, 45–47
*Animal's Agenda* (periodical), 6
animal welfare, 50, 53–54, 63
Animal Welfare Act (AWA), 17, 32
anitbiotics, 26, 34, 37, 43
anthrax, 53, 58
antiviral drugs, 22–24, 26
apes, great
  emotions exhibited by, 8
  language capabilities of, 20
  legal rights protection and, 9, 52, 53
  suffering of, during tesing, 29
  see also specific species
arsenic, 66
aspirin, 56
atrazine (pesticide), 62–63
autopsies, 27, 61

baboons, 46, 47, 48, 49
Baby Fae, 46, 49
Barnes, Donald, 33
biochemical pathways, 59, 61
bioethicists, 47
biological gap, 18–19, 65–66
biomedical research
  NIH study on government funding for, 41–42
  see also animal experimentation
biotechnology, 35, 38, 51
bioterrorism, 53, 58
birth defects, 37–38, 62–63
blood pressure, 26, 34–35, 36
blood transfusion, 59
Bogle, Rick, 8, 9
Bovine Spongiform Encephalopathy (BSE), 53
brain and brain function
  human vs. nonhuman primate
    differences in, 59
    similarities in, 7, 53
  nonhuman primate testing on
    futility of, 59
    as vital, 52